PRAISE FOR *LIVING IN THE LONG EMERGENCY*

"You won't find a better, more concise summary of what's really happening, the predicaments we face, and real-life examples of how ordinary people are responding. Optimism for the future begins with the awareness that things cannot continue as they have been. This book jumps that hurdle, and explores the past, the present, and the future in a way that is ultimately and surprisingly optimistic."

—Chris Martenson, author of *The Crash Course* and blogger at Peakprosperity.com

"Kunstler possesses the alchemy of describing a comprehensive disaster with a light touch. This is that rare, book on the future that is entertaining to the last page. The impression is that, along with the troubles, a more pleasant way to live will gradually emerge."

—Andres Duany, author of *Suburban Nation*

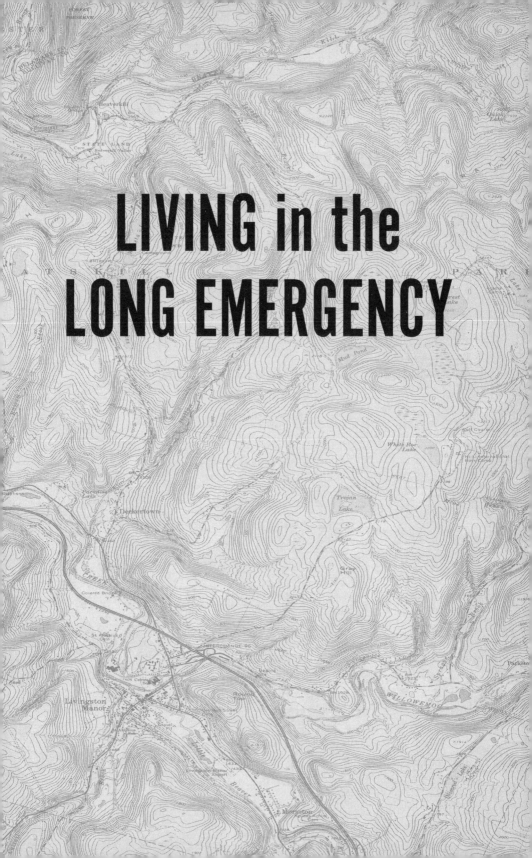

LIVING in the
LONG EMERGENCY

LIVING in the LONG EMERGENCY

Global Crisis, the Failure of the Futurists, and the Early Adapters Who Are Showing Us the Way Forward

JAMES HOWARD KUNSTLER

BenBella Books, Inc.
Dallas, TX

BenBella Books, Inc.
10440 N. Central Expressway, Suite 800
Dallas, TX 75231
www.benbellabooks.com
Send feedback to feedback@benbellabooks.com

BenBella is a federally registered trademark.

Printed in the United States of America
10 9 8 7 6 5 4 3 2 1

Library of Congress Control Number: 2019037850
9781948836937 (trade cloth)
9781950665129 (electronic)

Editing by Alexa Stevenson
Copyediting by Scott Calamar
Proofreading by Sarah Vostok and Amy Zarkos
Indexing by WordCo Indexing Services
Text design and composition by Aaron Edmiston
Cover design by Sarah Avinger
Printed by Lake Book Manufacturing

Map image courtesy of the U.S. Geological Survey (USGS)

Images on pages 7 and 10 by Steve St. Angelo

Distributed to the trade by Two Rivers Distribution, an Ingram brand
www.tworiversdistribution.com

Special discounts for bulk sales are available.
Please contact bulkorders@benbellabooks.com.

This book is for Steve and Joleen Meines.

"*We spend our lives trying to discern where we end and the rest of the world begins. There is a strange and sorrowful loneliness to this, to being a creature that carries its fragile sense of self in a bag of skin on an endless pilgrimage to some promised land of belonging. We are willing to erect many defenses to hedge against that loneliness and fortress our fragility. But every once in a while, we encounter another such creature who reminds us with the sweetness of persistent yet undemanding affection that we need not walk alone.*"

— Maria Popova

CONTENTS

INTRODUCTION

Back in the year 2005, I published a book called *The Long Emergency* that made the case for a coming collapse of the industrial economy. Since it predicted the demise of just about everything we consider normal in daily life, it spooked a lot of people. Here we are, fifteen years later. The country has seen the stunning election of our first black president, an epic financial blowup (and a dubious "recovery"), and the political shock of Donald Trump's 2016 victory. Yet to the casual observer it seems that little has really changed. The Ford F-150 pickup trucks still hurtle proudly around the ever-more-sprawling suburbs; the Too-Big-to-Fail banks still thrive in their artificial interest-rate arbitrage nirvana; the supermarket shelves groan with high-fructose corn syrup–based treats; Disney World rakes in record revenues; American troops still patrol the backcountry in Afghanistan; Silicon Valley keeps minting new billionaires; and, well, the whole wicked, groaning apparatus of modernity appears to carry on as if nothing significant has happened. It kind of reminds me of what Ricky Ricardo used to tell Lucy on TV: "You got some 'splainin' to do!"

All right, then, I will. For one thing, I didn't call it *The **Long** Emergency* for no reason. The operations of complex societies have many interesting features. Two in particular exist in a sort of dynamic tension of opposites: fragility and inertia. Fragility accretes insidiously as ever-greater complexity is layered onto the system. But inertia is the property by which systems in motion tend to remain in motion. A system as large and complex as ours has acquired tremendous momentum, which, of course, feeds back to aggravate its fragility, portending a more destructive eventual outcome. And so it keeps staggering along, despite all the tension and stress, until it reaches criticality . . . and cracks. And this can go on longer than we might suppose. Herb

1

Stein, the chairman of the Council of Economic Advisers long ago under presidents Nixon and Ford, summed it up nicely in Stein's Law: "If something cannot go on forever, it will stop."

In the so-called Great Financial Crisis of 2008, when fragility suddenly asserted itself under the weight of unprecedented mortgage fraud, Lehman Brothers collapsed, and a colossal daisy chain of bank counterparty obligations began to unravel, it looked for a while like the proverbial *end of the world as we knew it*. The cost to arrest this fiasco, including all the bailouts plus lost household wealth, ranged between $17 and $30 trillion, depending on who you ask. There will never be a credible accounting for it, but to paraphrase the late senator Everett Dirksen (R-Ill): *a trillion here, a trillion there, sooner or later you're talking about real money*.

The whole global system had been affected. In the aftermath, Mario Draghi, chief of the European Central Bank, said he would do "whatever it takes" to keep the international banking system—and hence the global economy—chugging on. He was obviously speaking for the whole central banking *community*. And so, since that time, the central bank financial wizards have executed every stratagem conceivable to maintain the appearance of stability, even while rot and failure spread from the margins to the center of civilized life. In this Potemkin economy, stock markets soared while the middle classes fell into an abyss.[1]

Yogi Berra's famous dictum, "It's tough to make predictions, especially about the future," beats a path to the Nobel Prize–winning work of psychologists Daniel Kahneman and Amos Tversky in the abstruse field of probability, uncertainty, and decision-making. One of their early experiments in the 1970s led them to conclude that human beings, given a little information, made worse predictions than people who had been given no information at all. Tversky later quipped, "The difference between being very smart and very foolish is often very small."

In writing about the prospects for our industrial civilization, I did what I could using my own applied heuristics (i.e., educated guesswork) to suss out

---

1 Potemkin economy: Named after Grigory Potemkin, minister under Catherine the Great of Russia, who, in 1787, set up false-front villages along the banks of the Dnieper River to impress the empress and foreign dignitaries as they toured past on a barge.

where our society was heading. I didn't have a lab, or research assistants, or an accepted corpus of mathematical models to reference. Anyway, hardly anybody in the American thinking class was paying attention to this story, and many of my observations seemed to take even educated readers by surprise, especially the news that we were heading into trouble with our oil supply. I concluded that the years ahead would bring an epochal discontinuity of economy and culture.[2]

In 2012, I published another book: *Too Much Magic: Wishful Thinking, Technology, and the Fate of the Nation*. After the 2008–2009 Great Financial Crisis, a species of wishful thinking that resembled a primitive *cargo cult* gripped the technocratic class, awaiting magical rescue remedies that promised to extend the regime of Happy Motoring, consumerism, and suburbia that makes up the armature of "normal" life in the USA. The nation wanted very badly to believe that all our accustomed comforts and conveniences would remain in place. This strange rapture provided reassurance that gee-wiz technology and sheer human ingenuity would propel us into an alt.energy-powered, computer-mediated utopia where old-school natural resource limits no longer applied to economics. What I called "the master wish" in this religion of wishful thinking went something like this: *Please, Lord, let us keep driving to Walmart forever.*

That national mind-set seemed neurotic and dangerous to me because the touted "recovery" was almost entirely a result of financial hocus-pocus on the part of governments and their central banks. Papering over the problem of insolvent banks and bankrupt governments with dishonest policies such as near-zero interest rates, printing money ("quantitative easing"), attempting to cure over-indebtedness with more debt, and other sneaky market interventions was sure to eventually lead to even more severe tribulations. The biggest mental block among the people running things was the fatuous belief

---

2 The same year *The Long Emergency* came out, 2005, also saw the publication of David Goodstein's *Out of Gas: The End of the Age of Oil* and Richard Heinberg's *The Party's Over*, as well as Dmitry Orlov's astute essay on now discontinued EnergyBulletin.net (now archived at Resilience.org), "Closing the Collapse Gap." Kenneth S. Deffeyes's *Hubbert's Peak: The Impending World Oil Shortage*, had been released in 2001 and his follow-up, *Beyond Oil: The View from Hubbert's Peak* came out in 2005.

in infinite industrial growth on a finite planet, an idea so powerfully foolish that it should have obviated their standing as technocrats. Interestingly, in this historical moment of maximum wishful thinking, nobody wanted to read a critique of wishful thinking.

The election of Donald Trump in 2016 shifted the public's attention from wishful fantasies about the robotic future of perpetual leisure and limitless wealth to the suddenly horrifying realm of populist politics. Trump made the thinking classes in America extremely nervous about the survival of the republic per se and many of its cherished moving parts, such as the two major political parties, the courts, the credibility of executive agencies like the Department of Justice and the FBI, and, of course, the role of the president. The nation was already disgusted with Congress well before the Trump-Hillary contest. In any case, after 2016, the attention of the media especially turned away from everything but the political battle of the so-called "resistance" versus Trump. There's very little news about climate change, the oil predicament, the condition of the banking system, the global food supply, the mass extinction of life on land and sea, and all the other issues that will likely end up mattering a whole lot more than the fate of Donald J. Trump.

Here's what I propose to do in this book: In Part One, I'll venture to discuss what I got right and what I got wrong making my prognostications in the earlier books. In Part Two, I'll present portraits of people I've met around the country who have been affected by the early stages of the long emergency, some of them battered by loss, some of them caught up in crafting new ways to thrive in the discontinuities yet to come. In Part Three, I'll try to answer the question: *Now what?* The objective is to stimulate the formation of a coherent consensus about what is happening to us so we can make coherent plans about what to do.

PART ONE

WHERE ARE WE IN THE STORY?

Chapter 1

HEY, WHAT HAPPENED TO PEAK OIL?

Early in 2018, the US Energy Information Administration (EIA) reported that this country had amazingly surpassed the old 1970 crude-oil production peak of more than ten million barrels a day. Quite a feat! By spring of 2019, it was up to twelve million. And look how rapidly it shot up after 2008, the year of the Great Financial Crisis. It took many decades to squiggle up that earlier 1970 peak, but this new move was more like a pole vault.

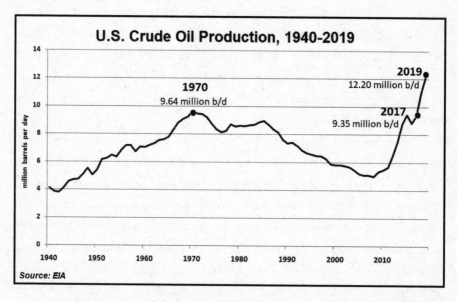

You can hardly blame the rank-and-file American public for being confused and complacent about where we stand in the oil picture—though news reporters, industry executives, and government officials surely ought to know better. Peak Oil didn't go away; it was postponed by a decade or so. In the meantime, the American public was bamboozled into thinking that we have no problem with our oil supply, that we have become "energy independent," a "net oil exporter"—and we can now look forward to a seamless transition into a post-fossil-fuel economy that will allow us to maintain our current standard of living just as it is, Happy Motoring and all, on other forms of energy, especially "renewables."

It's true that I did not anticipate the so-called shale oil "miracle," and it is not much consolation that neither did any of the other well-informed commentators who covered the Peak Oil story around the turn of the millennium, including distinguished petroleum scientists Colin Campbell, Jean Laherrére, Kjell Aleklett, Kenneth Deffeyes, as well as Caltech physicist David Goodstein, ecojournalist Richard Heinberg, and the late Matthew Simmons, all authors of books and monographs on the subject of Peak Oil.

A lot of the credit for the shale oil "miracle" goes to the Federal Reserve and Wall Street. The former kept interest rates supernaturally low for a decade after the 2008 Great Financial Crisis, which made for extraordinarily cheap financing. That provoked a feeding frenzy for borrowers and a frantic search for *yield* (return on investment) among bond holders. The Fed also pitched in with all the other major central banks, in an act of coordinated rotation, to prestidigitate those aforementioned trillions in new "money from thin air" to bail out the bankrupt global banking system (and rescue the investor class from its own poor choices).

Thanks to Wall Street, after 2008 quite a bit of this new money, along with those supernaturally low interest rates, was marshaled into "junk" bonds and other forms of high-risk borrowing that allowed the shale oil companies to pretend that shale oil was a viable business. Under a *normal* interest rate regime, say, of the mid-1990s when the benchmark ten-year US Treasury bond interest rate stood around 6 percent, junk-bond yields easily exceeded 10 percent. That's a lot of interest for an oil company to have to pay back on highly speculative drilling operations, at $6 million per well (minimum). In the *abnormal* interest-rate regime after 2008, junk-bond yields were only marginally higher than "risk-free" investment-grade bonds, such as US Treasurys'.

The trouble with artificially manipulated interest rates is that they misprice the true cost of borrowing money and lead to misinvestment in economically dubious ventures. In doing so, they artificially magnify booms and busts.

The shale oil "miracle," therefore, was a very impressive financial and technological stunt. In practical terms, it provided a means to pull forward from the future the last dregs of recoverable oil, so the US could *live large* for a few years longer. As independent oil analyst Arthur Berman put it: "Shale is a retirement party for the oil industry."

When *The Long Emergency* was published in 2005, the US was producing 5,187,000 barrels of oil a day, down from 9,640,000 in the peak production year of 1970.[3] By 2008, production had sunk even further to 4,998,000, down by half of the 1970 peak, on the face of it a rather shocking number. The model many of us followed then was Hubbert's curve, named after M. King Hubbert (1903–1989), the geologist and geophysicist who had worked for the Shell Oil Company for two decades and taught at Columbia University and Stanford. In 1956, Hubbert had proposed that the rate of oil production would follow a bell-curve pattern and predicted that American oil production would peak somewhere between 1965 and 1970. He was roundly criticized and denounced, especially within the oil industry, but he proved to be correct on that call. The year 1970 was indeed a US oil production peak. It wasn't evident or acknowledged until several years later—in the "rear-view mirror"—when post-1970 yearly production figures came in and showed that indeed the volume of US oil production had gone down for more than one year. The trend was for real.

The US production decline was arrested briefly between 1977 and 1985 when the Alaska Prudhoe Bay oil fields came on line, forming a lower, secondary peak at 8,971,000 barrels a day. (The substantial North Sea and Siberian oil fields also came on line in that period.) But after 1986, US oil-production declines resumed in earnest all the way to the year 2008, when production fell back below 1940s levels. However, US oil consumption remained relatively stable compared to the wilder swings in production and price.

---

3 US Department of Energy, Energy Information Agency (US EIA), https://www.eia.gov/dnav/pet/hist/LeafHandler.ashx?n=PET&s=MCRFPUS2&f=A.

Consumption is a different story. In 2006, US oil consumption hit an all-time high of 20,802,000 barrels a day. The low, following the Great Financial Crisis, only fell to 18,490,000 in 2009. So, we were still burning through enormous quantities, quite out of line with what we could supply for ourselves. Of course, it was necessary to import the difference between domestic consumption and production. Between 2005 and 2008, US imports reached almost fifteen million barrels a day—three-quarters of all the oil we burned. Our dependence on oil produced by foreign nations had become a matter of grave concern.

The oil price action (see chart below) was extremely erratic, rising steadily from $20 a barrel in 2002 to nearly $75 in 2006, then shooting up violently to $140 in 2008, and then plunging back to $34 shortly after in 2009, with the Great Financial Crisis. The price started bouncing right back up in 2009 and climbed back over $100 a barrel by 2011. It fibrillated around the $100 level until 2014, when it started to plunge again, back to flirting with the $30 and $40 levels until it commenced another decisive rise above $65 in early 2018, falling back to under $60 levels in the summer of 2019. The "bumpy plateau" behavior that I (and other reporters) predicted back around 2005 turned out to be even bumpier than we expected.

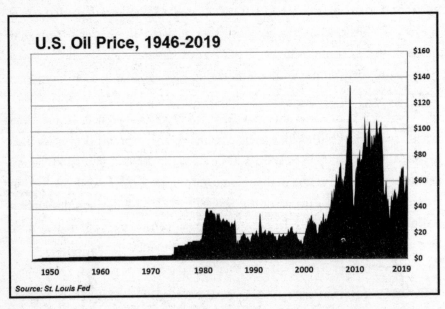

U.S. Oil Price, 1946-2019

Source: St. Louis Fed

I've been referring here to what the industry calls "conventional oil," which was mostly what was produced before 2005, that is, before shale oil came on the scene. Conventional oil is the type of oil you get from drilling a vertical hole somewhere and sticking a pipe in the ground like a soda straw. At first, the oil gushes out under its own pressure from natural gas in the source rock. After a while, the pressure gives out and the familiar "nodding donkey" pump jack pulls up the bulk of the oil in place without any additional interventions. It keeps pumping year after year, often for decades. The process requires no hydraulic fracturing, no convoys of water and sand trucks, no $600 million deep-water floating platforms. Technicians can easily drive to work and do the work in good weather, and the oil is transported over land by rail, truck, or pipeline to major collection hubs and refineries.

Conventional oil was what the twentieth century ran on and why it was a time of such dynamic economic abundance, boosting the US standard of living to new heights of comfort and convenience. In the 1950s, this kind of oil could be produced with a fantastic energy return on investment (EROI) of around one hundred to one. That is, for every one barrel's worth of energy you put into the process of pulling it out of the ground, you got one hundred barrels of "black gold" back. Conventional oil was therefore "cheap oil." Just about everything in the "developed" world was engineered to run on cheap oil, say $20 a barrel in today's dollars. Problems started to appear when the world's supply of conventional "cheap" oil peaked around 2005. Global consumption was still rising around the world. Increasingly, the oil we could get was not cheap. And not conventional. The EROI on *all* oil worldwide was falling from that glorious one hundred to one ratio in the 1950s to more like fifteen to one. The EROI for shale oil was a dismal five to one. Tar sands were even worse.

Another big issue was that the discovery of any kind of new oil had been trending down since as far back as the 1960s. We were just not replacing the volumes of oil that we burned year after year, and you have to find oil before you can produce it. Most of the great finds of the oil age—the "elephant" fields of Arabia, Russia, Texas, Alaska, the North Sea, Mexico, Venezuela, Iran, Kazakhstan, et cetera, had been tapped—and nothing new on that scale was turning up. The shallow-water offshore fields in the Gulf of Mexico were all mapped. The "low-hanging fruit" had all been picked.

Exploration was increasingly directed to regions of the world where it was more difficult to drill for oil, notably deep water, and to new and extremely complex technologies for getting it out of the ground. The capital expense was enormous, and the hazards were great. Drilling miles under the water and the rock beneath it, where the business end of the drilling rig met powerful geologic forces far from human control, could produce stupendous disasters, like the 2010 BP Deepwater Horizon blowout, which fouled the Gulf Coast and cost $61 billion to clean up. In any case, there were long lag times between discovery of oil and actually getting the rigs up and running. And drilling dry holes—where prospects just didn't pan out—also became financially punishing under these circumstances. In short, the newer oil just didn't pay off the way the old oil did. Discovery of new oil hit a seventy-five-year low in 2019, preceded by several previous record low years.

By the early twenty-first century, the picture for new oil was growing rather grim, and the peak of conventional production around 2005 ended up rocking the financial world, though it took a few more years to manifest in the Great Financial Crisis. Rising oil prices tended to depress economic activity for the obvious reason that petroleum played a role in practically everything we produced or every service we provided. Since there were fewer places to even bother searching for new oil deposits, the industry turned back to the oil fields that were already known. Some of them had considerable oil still left in them, but it was trapped in shale formations that had *low permeability*, which referred to the ability of fluids to move through the rock.

The old conventional oil came out of *highly permeable* porous rock such as sandstone, which was spongy. That oil flowed through it easily to the wellhead, first propelled by pressurized methane gases that accompany oil, and then sucked out by the nodding donkey pump jacks. In the shale formations, oil couldn't flow through the rock. It was trapped; you had to blast it out. That's where hydraulic fracturing (i.e., "fracking") came in. If you could get a pipe into the shale strata, you could blast it with water under high pressure, and then inject sand along with the water to hold open the tiny fractures. Oil would seep out of the fractures and you were back in business.

The good thing was that you could do this on dry land in places where, at least part of the year, the weather was fine and the logistics were favorable for moving people and equipment onto the site. The Bakken shale play, straddling North Dakota and Montana, had been worked for decades with regular vertical holes, as were the Eagle Ford play and the Permian Basin in Texas. They were old and tired. Fracking held out great promise to an industry that was, otherwise, running out of options. The downside was that shale required a lot more capital investment than the old conventional oil. A shale oil well, including fracking operations, might cost $6 million (in current dollars), as compared to $400,000 (in current dollars) for an "old school" East Texas–type well of the 1950s.

Another downer was that shale oil wells had low daily flows and very rapid depletion rates. Shale wells typically decline from 70 to 90 percent after just three years. The average daily well flow in the Bakken region was under one hundred barrels a day in 2018, only a little better than the old school "stripper" wells, as played-out conventional wells on their last legs were called. By way of comparison, daily flows of old-school conventional wells in their prime were *thousands* of barrels a day, and they continued at that rate for decades. Considering the low cost of old-school drilling and pumping, these wells were like cash registers.

Fracking oil-bearing rock with explosives goes all the way back to the nineteenth century. It was used early on to tease out some extra production in played-out conventional wells. But the fracking process as we know it today, using high pressure water injections and sand rather than nitroglycerine, got going in the 1990s. The game-changer was horizontal drilling, which, aided by computer-mapping technologies, enabled the oil companies to work the thin layers of tight shale strata with extraordinary precision. The well bore might go a mile or more down vertically, and then make a forty-five-degree turn and run horizontally for another mile into a shale layer a few meters thick. All that had to be fitted with pipe and cemented in place. Then, massive amounts of water and sand had to be trucked to the site for fracking operations. (It had to be a special kind of sand, too: high-purity quartz with very uniform, round grains—an increasingly scarce and expensive commodity.) Much of the frack water (seven to fifteen million barrels per well, plus an array of chemicals to enhance the process) came back up with the oil. It then had to be separated and trucked back off-site for disposal—all of this representing considerable additional capital costs.

On top of all that were the environmental "externalities"—the threat of poisoned groundwater from chemicals in the fracking fluid, seismic disturbances (i.e., earthquakes) from poking so many pipes through underground strata and blasting it with high-pressure water, the considerable wear and tear on roads from the cavalcades of water and sand trucks, and the higher-order damage to the biosphere from extending the fossil-fuel economic regime.

There was a lot of proud talk about constant improvement in the technology. True, the engineers got better and better at lengthening the lateral frack lines and increasing the rate of initial production. However, that only pulled forward the estimated ultimate recovery (EUR) for the life of the well. You got more oil initially, but it depleted sooner—kind of like the old gag where the guy wants to make his blanket longer, so he cuts a foot off the top and sews it onto the bottom. To just keep production flat (i.e., not falling), companies had to incessantly drill new wells (at $6 million per) and re-frack the older ones (more truckloads of water, sand, etc.). They were also tempted to space their wells closer together, ostensibly to keep flows up, but that, too, only led to self-cannibalizing their production and more rapid depletion.

The big shale oil basins were not uniformly productive. They had a limited number of "sweet spots," and naturally these were the sites exploited first. That being the case, the supply of sweet spots dwindled over the years of the shale boom. Production in the Bakken and Eagle Ford plays began to fall off in 2014 and now look like they have entered terminal decline. By 2018, the Permian Basin was where the action had shifted. For the time being, it is the region of greatest oil production in the USA. It's a larger region than either the Bakken or Eagle Ford, but there's no reason to believe that the sweet spots are a whole lot greater and sweeter there. And the oil companies are under tremendous pressure to produce as much oil as quickly as possible in order to maintain the cash flow necessary to service their debt. Paradoxically, that has led to excessive production and thus lower oil prices—a hamster wheel of futility.

The official language of the oil industry and the government for predicting how much oil is in place anywhere can be very fudgy. The words "resource" and "reserves" can mean very different things, which leads to much confusion in the news media and, ultimately, among the broad American public. As analyst Robert Rapier wrote:

> "resource" means that they are estimating the technically recoverable oil in place. This says nothing of the economics of recovering this oil. The amount that would be economically worthwhile to recover at prevailing commodity prices—which would be classified as "proved reserves"—will be a smaller subset of the assessed amount. It would even be zero at a sufficiently low oil price.[4]

Obviously, the equation for getting this oil is a lot different at $120 a barrel than it is at $30 a barrel, two hash marks that the industry has swung between in the past decade—though even at the high end of the price spectrum, shale oil production was not turning a net profit; the capital borrowings were colossal. The US Geological Service survey estimated that the Permian Basin contained somewhere between eleven billion and thirty billion barrels of technically recoverable oil. The US burned around five billion barrels of oil in 2017 plus an additional two billion barrels of natural gas liquids, averaging around 19.5 million barrels a day. Independent analyst Arthur Berman estimated that the Permian Basin contained a remaining 3.7 billion barrels of realistically economical-recoverable oil. Do the math. "I imagine that may surprise many who buy into the vision of American energy dominance," he remarked.

The claims ballyhooed in the news media, and reverberating through the echo chambers of the internet, that America is a net oil-exporting nation and "energy independent" are just plain untrue. The United States produces a large share of the petroleum it consumes these days, but it still relies on imports to meet demand. At this writing, US oil imports were just under

4 Robert Rapier, "The Permian Basin Keeps On Giving," *Forbes*, November 21, 2016, https://www.forbes.com/sites/rrapier/2016/11/21/the-permian-basin-keeps-on-giving/#31b3d0dd4a2a.

four million barrels a day (out of the roughly nineteen million barrels a day consumed).

Not all crude oil is the same—or equally valuable. Shale oil is an exceptionally light grade of oil. It contains about 90 percent of the energy of, say, Saudi Arabian crude, or the Oklahoma crude of yore. It comes out of the frack pipe almost as light as gasoline. It doesn't yield much middle-grade transportation fuels such as diesel and aviation fuel. And, significantly, US refineries are not engineered to process it. We mix some of it at the refineries with heavier grades of conventional crude, but there's only so much that can be used. Some of it is sent to other countries to mix with heavier crudes they produce, and a lot of that is sold at a discount because diesel engines predominate in European cars and shale oil doesn't yield much of those heavier distillates. Some of US shale oil ends up being exported as "finished product," i.e., gasoline, because that is the main distillate in shale oil, and in the midst of the current shale boom, we're producing more gasoline than we can use domestically.

Why don't US oil companies build new refineries that can handle light shale oil? The fact that they're not doing this tells us something: namely, that the enormous investment to do so would take the thirty-year lifetime of a refinery to amortize financially, and the industry knows that shale oil production has a much shorter life expectancy than thirty years. Based on the history of the declining Bakken and Eagle Ford plays, the Permian Basin, too, may be in decline within a few years. In short, the oil companies recognize that they're in a sunset industry (Art Berman's "retirement party"). In the meantime, they are burning through cash. From the start of the shale oil boom a decade ago, the companies' operating expenses have exceeded their revenues at any price. The ultra-cheap cost of borrowing tended to conceal that because they could always fill their cash coffers with new borrowings. The rest was accounting malarkey. Shale oil just could not pay back the costs of production. All the companies' cash flows went to paying down their borrowing costs, and after that they faced actual net losses.

The US is caught in a predicament that can be stated pretty simply: oil over $75 a barrel crushes economic activity, and oil under $75 a barrel bankrupts oil companies. The price leaps of the early 2000s from $20 a barrel (what an advanced industrial economy could afford to run on) to over $150 in 2008 substantially damaged the US economy in ways that have never been fully accounted for in the news media. It was certainly a major factor in the Great Financial Crisis.

Afterward, the interventions of the Federal Reserve succeeded only in blowing investment bubbles of every kind that, in essence, represented ever more borrowing from the future to compensate for the lack of the affordable primary resource (oil) needed to run everything. After the year 2000, US economic "growth" was concentrated in two sectors especially: financial engineering on Wall Street and construction of sprawl housing in the outermost exurbs. These activities were related, since the products of the housing bubble included billions in securities composed of janky home mortgages, the notorious *collateralized debt obligations* (CDOs) that blew up in 2008 and had to be ring-fenced on the Federal Reserve's balance sheet, where some of this worthless paper resides to this day.

Then came the bailouts, three rounds of quantitative easing (money printing), Operation Twist (fiddling the ratio of Fed purchases of short- and long-term bonds), ZIRP (zero interest-rate policy), and other officially sanctioned machinations since the financial fiasco of 2008 that amounted to tens of trillions of dollars. These kept the game going in the financial markets but could not conceal the fundamental fact that we can no longer afford to run an advanced technological society at the scale we do on the energy resources at hand. The final reckoning of this melodrama will begin when interest rates in the bond markets can no longer be suppressed. Rates will rise because the public—including investment professionals—will come to accept the reality that there is too much outstanding debt that will never be paid back. That is when the easy money stops flowing and the false economics of unconventional oil will fail.

NATURAL GAS ADDENDUM

Natural gas, a.k.a. methane, is mostly used for electric power generation, home heating, and the chemical industry (plastics, fertilizer, insecticides, and many other products). It is very important to the nation's well-being as an advanced economy. In 2005, natural gas production was in terminal decline. Production then came from old-school conventional gas wells, no fancy tricks, just a vertical pipe in the ground. A lot of it was associated with conventional oil wells. Back in the mid-twentieth century, it was considered to be of such marginal value that it was simply lit up and flared off. But over the decades, as coal-fired electric power plants were phased out, and nuclear power became expensive and problematic so that almost no new nuke plants were built, the USA turned increasingly to natural gas for growing electric-generating capacity. By 2005, the situation had become quite worrisome. Supply was falling, and the price was spiking up above fifteen dollars a unit (1,000 cubic feet expressed as *mcf*).

The fracking revolution turned that around so spectacularly that by 2013, the US was seeing new all-time highs in production, and unit prices had fallen between two dollars and three dollars, where it remains at this writing—a price that does not provide profits for the shale gas companies. The US EIA projects many decades of abundant supply.

We would do well to be skeptical. J. David Hughes publishes annual reviews of shale oil and gas reports (*Shale Reality Check*) that dispute the extremely rosy annual outlooks published by the US EIA. Hughes, a former research manager for thirty-two years at the Geological Survey of Canada, where he headed up research on unconventional gas and coal, maintains that production in the earliest shale gas plays—the Barnett, Haynesville, and Fayetteville—peaked in 2016 and has declined more than 40 percent since then. Prospects are good for the moment in the larger Marcellus shale gas play straddling Pennsylvania and West Virginia, but there, too, initial production was goosed as high as possible at the expense of future production, so the Marcellus is liable to see the same kind of steep falloff as the others by the early 2020s.

Hughes wrote of the EIA's 2017 Annual Energy Outlook:

The reference case projects that 1.29 million wells will be drilled to recover oil and gas . . . in the period 2015–2050. At $6 million per well, this amounts to

$7.7 trillion . . . Given the EIA's overestimates of future shale production and recoverable resources, it is unlikely that all of these wells will be drilled.[5]

Hughes is not alone in his thinking. His conclusions were ratified by forecasts put out by the Bureau of Economic Geology at the University of Texas, Austin, which predicted that the four biggest plays in the US would peak by 2020, then decline, and eventually produce only about half as much as the EIA's forecast.[6]

Given that field declines are steep, requiring 25–50 percent of production to be replaced each year, the levels of drilling and capital investment needed to maintain production will escalate going forward. Yet the USA doesn't have $7.7 trillion to continue this noneconomic (i.e., profitless) endeavor, even in an interest-rate regime that remains relatively very low as of 2019. Rising interest rates would make the $7.7 trillion absolutely inconceivable without new rounds of money printing by the Fed—which would be very damaging to the US dollar, with unappetizing ramifications, including a higher cost for the large amounts of oil we import each day.

The benefit of the shale oil and gas "miracle" was that it postponed the reckoning over these resources by about a decade. But it was achieved at the cost of badly misleading the public about the eventual crisis, and by probably making it worse when it gets here.

The public apparently has lost any sense of urgency about our fossil-fuel quandary, and therefore lost any motivation to even think about making changes in our basic living arrangements. The project of suburban-sprawl building has continued, albeit not at the same pace as pre-2008 due to the ongoing impoverishment of the middle class and the debt burdens of millennials entering their household formation years.

---

5 David Hughes, *Shale Reality Check: Drilling Deeper into the EIA's Rosy Projections for Shale Gas & Tight Oil Production Through 2050*, 2017, http://www.postcarbon.org/publications/shale-reality-check/.

6 Nafeez Ahmed, "US shale boom has already 'peaked' says former govt geoscientist," *Insurge Intelligence*, December 15, 2016, https://medium.com/insurge-intelligence/us-shale-boom-has-peaked-says-former-govt-geoscientist-4c2d23c7412f.

Instead, the public has become convinced via propaganda from the high-tech sector and the car industry that we will simply replace gasoline-powered cars with electric cars, including automated self-driving electric cars in continuous circulation like taxicabs, making possible, in theory, a suburban-sprawl landscape that can be serviced by fewer cars altogether. I'll discuss these hopes and expectations in the next chapter, along with other alternative-energy schemes currently afoot.

Chapter 2

THE ALT-ENERGY FREAK SHOW

At least Elon Musk has a sense of humor. The entrepreneur made a fortune creating PayPal; went on to found Tesla, Inc., the electric motorcar and battery-storage company; and recently started a private space exploration company called SpaceX. He test-fired his Falcon Heavy rocket in early 2018, launching a Tesla car into orbit with a mannequin in a space suit named "Starman" strapped into the driver's seat. The car was a Roadster convertible with the top down. Nice ride. A readout on the dashboard screen says "Don't Panic!"

But Mr. Musk is not kidding with his various grandiose ambitions. He has taken up the baton of an exhausted NASA with the "aspirational goal" of colonizing Mars. "The future is vastly more exciting and interesting if we're a space-faring civilization and a multi-planetary species than if we're not," he explained.

Well, I'm not so sure. Unless, that is, we're tired of the possibilities for joy, meaning, and excitement on a planet (this one: Earth) that we are superbly fitted to thrive on—and which, sadly, we're in the process of damaging quite recklessly with our current activities, including shooting a lot of junk into orbit around it. Aside from the sheer excitement of space travel, Mr. Musk explained his Mars colonization venture as setting up "a backup hard drive" for humanity, in the event that an asteroid bashes into the planet or we go through another episode of mass extinction, like the five previous ones in Earth's history, or some other calamity ensues to threaten the human project here. The fantasy of escaping Earth for some alt dwelling place in the heavens is hardly a new thing for the human race, with its bent for mythmaking.

Considering the scary problems already upon us, from climate change to the death of the oceans to population overshoot, and much, much more—and now that techno-narcissism has replaced religion for expressing our vested metaphysical hopes and dreams—Mr. Musk's Martian adventure has special allure for the educated classes these days.

But the Mars colonization proposal does raise a sticky question: If the human race can't get its shit together on Earth, how might we possibly thrive on a distant planet with an atmosphere that is 95 percent carbon dioxide, has little protection against space radiation, no visible aboveground water, nor any of the geophysical and biological characteristics that support life here, and which, finally, is a 33.9-million-mile resupply journey at its closest? (Earth and Mars both follow elliptical orbits, with the *average* distance between the two planets at 140 million miles.) Kind of difficult to send out for spare parts or a few tons of pepperoni sticks.

If space exploration is Mr. Musk's avocation, his main business has been the production of electric cars and development of improved electric battery power storage. The intense wish among the broad American public to "solve" our perceived problems of fossil fuels and climate abnormalities by electrifying the US auto fleet (of around 260 million passenger vehicles) is one of the more quixotic expressions of the national neurosis that has been our response to the long emergency.

For one thing, it assumes that we can continue living in the suburban sprawl settlement pattern, which I consider an unrealistic and impractical proposition. The wish to mitigate the predicament of being stuck living in suburbia may be understandable, given our vast investments over four generations in that colossal infrastructure, and given how familiar and comfortable it is. But that does not make it any less impractical. The last thing you hear from advocates of electric mass motoring is any advocacy for walkable, human-scaled communities, which is by far a more realistic and practical remedy for the fiasco of suburbia. And, of course, there is close to zero interest in reviving the US passenger train network to connect walkable towns and cities. We've probably missed the opportunity for that anyway, because of the relationship between our dubious energy supply and the coming impairments capital formation associated with our energy quandary.

The electric car advocates assume that the electricity for these vehicles will be produced with no harmful externalities. That overlooks the fact that so much of the electric power available on the grid comes from coal, natural gas, and nukes, which, in one way or another, all produce emissions and toxic by-products. So that when you are tooling down the highway in your emission-free Tesla car, be advised that some power station a hundred miles away is emitting for you.

Additionally, the electric car fleet discussion generally bypasses the energy embedded in the manufacturing process required to make each car. The percentage of US electric power sourced from wind and solar went from inconsequential in 2005 to just under 5 percent in 2017, obviously an improvement, but perhaps a little ominous considering our dependence on depleting shale gas (31 percent of total electric power production) and aging nuclear (20 percent), with most US reactors already past or near the end of their design life, and very few ones underway to replace them. Coal, by the way, powers about 31 percent of our electric power and the rest is hydro—that is, turbines powered by dammed rivers.

There is plenty of talk about reconfiguring the extant electric power grid—the world's biggest machine, so to speak—into a "distributed network" arrangement of innumerable local solar and wind installations hooked together, but so far that has not happened. Adding additional loads of millions of proposed new electric cars to the US power grid doesn't really work with the situation I have just described. The growth curve of solar and wind would have to steepen very quickly in the decade ahead. Even if astonishing improvements in photovoltaics come on the scene, the worsening financial picture will be an obstacle to reconfiguring US electric service.

The fantasy that power for the proposed electric car fleet will someday come from "renewables" also probably founders on the unacknowledged need for an underlying cheap fossil-fuel economy to fabricate things like solar panels and wind turbines at the necessary scale to come close to the lifestyle we think of as normal. It rests on assumptions about available capital that don't comport with the reality of impaired capital formation in a non-cheap-energy economic environment. If energy is not affordable—with $20 a barrel oil as the benchmark—our techno-industrial economy will not produce surplus wealth, which is what capital is. And oil has not been that cheap for nearly twenty years. Compensating for the inability to generate real surplus wealth (savings) by having the central bank print money, as we have done since 2008, has distorted the functioning of financial markets and can lead eventually to the loss of faith in money itself, a process well under-way. The failures of capital formation have other profound ramifications that are already upon us.

For instance, the incremental impoverishment of the former middle class is seen these days in the difficulties they have qualifying for car loans—which is how Americans are used to buying cars—so lenders have bent over back-ward to shoehorn desperate customers into new extremes of loan arrange-ments. New cars have also jumped in price. Since car buyers tend to focus on the monthly payment rather than the total price, the solution has been to extend the term of the loan to keep the monthly payment low. The payment figure is used to determine whether customers qualify for a loan, and even that has become a squishy criterion—so anxious are the dealers to move the merch off the lot. Interest rates were kept supernaturally low under the Federal Reserve's ZIRP and near-ZIRP policies, which also helped keep the system going.

Unfortunately, cars depreciate quickly, and the buyer with a seven-year loan can easily end up "underwater" as the years go by, with collateral (the car) that is not worth the remaining payments on the loan. Longer loan terms also generate more total interest paid over the life of the loan, bloat-ing the ultimate price of the car. Lately, we see subprime loans to sketchy borrowers that are not unlike the subprime mortgages that caused so much trouble in the housing market in 2008. Interest rates on these subprime loans can be 14 percent or more. Lenders are even offering seven-year loans for used cars, which are likely to be junkers well before the debt can be paid off.

To make matters worse, much of that sketchy debt has been securitized, repackaged into bonds—collateralized loan obligations (CLOs)—like the home-mortgage–based CDOs that crashed the system previously. These she-nanigans are a prelude to the coming scarcity of lending capital as it becomes clear that the global mountain of already-existing debt will never be paid back, making the issuance of new debt impossible. The result: Americans will increasingly be foreclosed from buying into the Happy Motoring experience—whatever way the cars are powered.

As this set of problems and impediments presented themselves in the years after the Great Financial Crisis of 2008, a new spin on the electric-car story was added with developments in artificial intelligence that, in theory, would make self-driving cars possible. Under that wished-for regime, it would no longer be necessary for Americans to own cars. They could simply summon a self-driving electric vehicle at will, sort of a universal taxi service, either on the business model pioneered by Uber, or some other corporate arrangement. This supposedly would lead to a much smaller national vehicle fleet while giving a new lease on life to suburbia. Of course, that raises a lot of interesting questions, too, such as . . . just how many self-driving cars would be required for the morning commute in places like Phoenix, Dallas, Atlanta, New Jersey, the San Francisco Bay Area, and other major urban metroplexes? Even if the trips were all multi-passenger rides, the commute would still require an awful lot of cars.

I saw a strange preview of that not so many years ago in Johannesburg, South Africa, a city the size of Phoenix in population. In the absence of a municipal bus or tram system, a colossal taxicab fleet transported masses of low-wage workers across the city twice a day from the slums of Soweto to the northern corporate suburbs, making for stupefying traffic congestion during commute times that extended for several hours at each end. And then, what did these additional vehicles do during the midday interval? They didn't pick up fares, because the better-off people in Jo'burg had their own cars. So, the taxi drivers hung out idly around giant car parks in the old downtown, where the skyscrapers had been abandoned by corporate tenants, who had moved into new fortified quarters in those sprawling burbs when apartheid ended. The cost of taking these taxis to and fro cost the poorer workers a considerable chunk of their paltry wages, as well as eating up at least a couple of hours at each end of their workday. In the imagined future driverless-car nirvana of the USA, there would be no midday socializing at the car park, obviously, but I suspect the logistics would otherwise be similar.

Then there is the question of how, in this Uberized driverless electric-car utopia, the car industry might survive making only a fraction of the vehicles it was tooled up for in the old days. The US automobile industry was selling around sixteen to seventeen million cars a year until the financial crises of 2008. Sales in 2009 fell sharply to below ten million, which provoked the bankruptcies of Chrysler and General Motors. They were bailed out by the federal government and restructured on a leaner basis with broad concessions from the United Auto Workers union. (Ford went its own way and somehow survived.) The Obama administration's "Cash-for-Clunkers" gimmick bumped up new car sales briefly in 2009. But then sales fell back again when the clunkers ran out. (A lot of those older cars were simply destroyed to prevent them from flooding the used car market.) After 2010, car sales crept back up again to over seventeen million a year. These are the kind of sales numbers the car manufacturers need to justify their business model. How does this gigantic industry reorganize for much-lower-scale production and survive? Does it become a boutique industry? I'm not persuaded that it can be.

Electric motors are far simpler than gasoline or diesel engines. Electric motors plus their gearing and differentials can be assembled with fewer than a hundred moving parts; internal combustion cars require more than a thousand. The good news would be that electric cars have the potential to last much longer than gasoline or diesel cars. That's good news for the car buyer, at least. The bad news is that, unless cars are engineered for timely failure (planned obsolescence), the car manufacturers cannot depend on the same routine replacement formula that has ruled their economic model for decades and accounts for most car sales. The developing picture looks like a kind of race between a floundering customer base and a floundering industry, with all the built-in terminal problems of the suburban living arrangement sandwiched between.

Elon Musk appears to have been motivated to do good for the world. Giant ego aside, let's give him the benefit of the doubt. His electric car company, however, is struggling. The first generation of cars was aimed at the luxury market. Tesla cars soon became the leading consumer status item in the few parts of America where people were still thriving, especially the two great poles of wealth in California: Silicon Valley and Hollywood. Beyond the obvious status signal that the Tesla owner is well-off, it offers great opportunities for environmental virtue signaling: "I'm among the elite greens! No tailpipe emissions here! Admire me!" The Model S retailed for just under $70,000 in 2018, clearly out of reach for the masses of corporate cubicle serfs and service industry workers (waiters, retail clerks, beauticians, lab technicians, etc.) who represent what's left of the middle class. The market for Tesla luxury models became saturated quickly.

In the spring of 2018, Mr. Musk was having a lot of trouble producing his Model 3 Tesla car. The vehicle was pitched as a Tesla for the masses, but with the price starting at $35,000, that seemed misleading. In any case, Mr. Musk had initially promised a production rate of 2,500 units a month when he introduced the car, but they were barely turning out 1,000 a month, and the rumor was that these were being assembled by hand, like 1923 Stutz Bearcats. In March of 2018, the rubber hit the road for Mr. Musk's company. Moody's downgraded the company's bonds, and the share price fell 30 percent. It burned through $500 million in cash in the fourth quarter of 2017 and was projected to burn through $900 million in each quarter of 2018.

In March 2018 a Tesla car navigating on autopilot crashed into a concrete median barrier in Mountain View (Silicon Valley), California. The car's battery burst into flame. The driver was killed. The same month, a self-driving Volvo car run by Uber struck and killed a woman walking her bicycle across a street in Tempe, Arizona—more bad press for the wished-for transformation of motoring. Of course, Americans long ago got comfortable with a mobility system that kills around 40,000 people annually. But the psychology of that works differently when you can blame a human being who makes an error behind the wheel as opposed to the incomprehensible workings of an onboard computer.

In the summer of 2018, Musk astonished the financial media with a nine-word tweet offering to take Tesla private—that it is, buy it back from the shareholders—at $420 a share. It was running around $370 at the time. That would have made for a purchase price of $72 billion. Musk said he had the financing for that all lined up. By late September, the Securities and Exchange Commission (SEC), which regulates the stock market, had fined Musk $20 million and forced him to resign as chairman of the board, though allowing him to remain as CEO. He was also required to clear any future tweet communiqués with the company's board of directors. The SEC determined that Musk did not, in fact, have the financing in place, as claimed, to do the buyback deal, and that the tweet had amounted to a manipulation of the share price. Some shareholders sued.

The company hit additional speed bumps that year. In June, it laid off 9 percent of its workforce. The solar roof tile segment of Tesla's portfolio—a big part of its hype—had only managed twelve installations by midyear. By October 2018, Tesla stock was down to $250 a share. Musk had already given up trying to take Tesla private. Its credit was downgraded by Moody's. There was turmoil in the upper levels of Tesla's management, complaints of workplace conditions in Tesla's factory, and several fires in the facilities. Musk himself asserted that the production line was overautomated with too many robots. "Humans are underrated," he told a *Wall Street Journal* reporter.

Many of the other major carmakers are tooled up for electric vehicles (EVs) now, but you sense their hearts are not in it. The price of these cars averages well over $30,000, which bumps us back into the issue of the financially struggling middle class. It's possible that the price of EVs will drop into some zone of affordability, as computers and flat-screen televisions have, but the middle class could lose ground even faster than EVs get to that affordability sweet spot. Hardly anybody is questioning the basic assumption behind the whole motoring quandary: that we must continue to be a car-dependent nation at all costs.

The likelihood that we will power the USA on "renewable" energy in anything remotely like the current configuration of activities—suburbia, Happy Motoring, air-conditioning for all, cheap food, night baseball, Netflix, Amazon, server farms, commercial aviation, et cetera—is about the same as the chance that Xi Jinping will deliver each and every one of us a dim sum birthday breakfast at home next year. The public is tragically confused about so-called "renewable energy." The sun may shine a lot of the time (in some places anyway), and the winds may blow (ditto), and they may indeed be eternal features of Earth's geophysics, but the hardware necessary to capture that energy is not renewable. It's a product of a fossil-fuel economy, and we have no experience fabricating this hardware any other way—most particularly via renewable energy sources. And certainly not at the scale required for a vaunted "Green New Deal."

Less has changed in this realm since 2005 than the reader may think, given the volume of propaganda about "innovative technology." Thirty years ago, 89 percent of the world's energy came from fossil fuels. Now, it's around 85 percent. A lot of the remaining 15 percent comes from nuke plants. Hydro is about 7 percent. About 2 percent is solar and wind. Not a giant step.

The wishful public has been fed a diet of misinformation from a wishful news media that won't tolerate anything but positive thinking about maintaining our current arrangements because imagining a different outcome is too depressing. This is not a malicious conspiracy by evil authorities so much as a neurotic defense mechanism in the face of the disturbing reality that the comforts and conveniences of recent decades may be drawing to a close. This national thought disturbance exists as an intransigent consensus that will resist revision until disruptive events—oil crises, war, electric blackouts—make the situation obvious and irrefutable.

Computer wizardry, especially the marvels of digital imagery that can make anything *seem* real on a pixel screen, have deceived the public into thinking that techno-magical rescue remedies are standing by just waiting to save advanced civilization from resource scarcity, climate disturbances, and financial disorder. After all, if we can make moving images as thrillingly and persuasively realistic as a *Jurassic Park* movie, how can we fail to run 130 million households on energy from the sun and wind—forces that we see and feel most every day and are just begging to be put to good use? The forces that we don't see as clearly as sun and wind are the forces of physics—the realities of the universe—especially as they are embodied in economics. These high-tech renewables just don't add up, and certainly not at the scale of the civilization we're living in. At a much more modest scale, using older, simpler technologies, there are many ways to harness wind and sun. But that is not what the debate has been about. For high tech, on the scale of places like Houston or Milwaukee, the math is unforgiving. Every element in the renewables picture costs more in sheer energy, or its representation as money, than we're able to put into it, or that we would be able to get out of it, or that we would need to maintain and repair the hardware.

There are many impressive demonstration projects already operating around the world, but it may be a sad fact that they will turn out, like shale oil, to be an impressive short-term stunt. Cheerleaders for renewables point to Denmark, which produces about 40 percent of its electric power with wind, as the poster child for the expected "green energy transition."[7] That is only possible because Denmark has power plants fueled by natural gas operating in the background to make up for the intermittency of the wind and balance its grid load. Often, the wind blows too much at the wrong times, and Denmark has to get neighboring countries to absorb the surplus they produce at giveaway prices, because there's nowhere to store it otherwise. At other times they have to import electricity from neighbors. Overall, Danes pay the highest electric rates in Europe.[8]

Giant wind turbines in huge wind-power "farms" would not be possible without fossil fuels. We've deployed many such installations the past decade or so, and at some future point these machines will have to be fixed or replaced. After all, they are subject to great forces of nature that wear them down, especially the wind turbines built offshore in corrosive seawater, which happens to be the case in Denmark. The manufacturers of wind turbines claim a useful life of twenty to twenty-five years. One study carried out by a team at Edinburgh University concluded that the productivity of wind farms they examined had degraded by one-third after ten years' service, making them questionably economical to operate.[9]

Wind turbines typically contain more than eight thousand parts. These are made of steel, concrete, exotic metals, and exotic plastics, components that depend on heavy mining activity, the petrochemical industry, long supply lines, and a lot of energy to bring it all together to manufacture and

---

7 Does not include fossil-fuel-powered cars, trucks, heating, manufacturing, or backup electric generation, just 40 percent of all electricity.

8 Alice Friedemann, "Wind and Solar diurnal and seasonal variations require energy storage," *Peak Energy & Resources, Climate Change, and the Preservation of Knowledge,* June 4, 2015, http://energyskeptic.com/2015/wind-and-solar-diurnal-and-seasonal-variations-require-energy-storage/.

9 "Wind turbines' lifespan far shorter than believed, study suggests," *The Courier,* December 29, 2012, https://www.thecourier.co.uk/news/scotland/82974/wind-turbines-lifespan-far-shorter-than-believed-study-suggests/.

then deploy the gigantic machines. Then there are significant operational and logistical costs, especially where the towers are placed offshore, including the difficulty of moving maintenance crews out to sea and back. From a climate change standpoint, it's debatable if all this embedded fossil-fuel-based activity negates the theoretical CO_2 savings of using wind turbines to make electricity.

I am skeptical that a lot of the materials needed to build wind turbines will even be available a decade or so into the future. Most of the rare earth metals used in the magnets at the heart of electric turbines are mined in China and Mongolia these days. (By the way, the process creates a great deal of toxic waste, some of it radioactive.) Geopolitical friction has increased since 2005, and global trade relations are showing the strains in tariffs, sanctions, and currency shenanigans, all tending to undermine the globally linked economy.

This sort of conflict historically has led to military mischief. I wouldn't bet on the continuation of benign international conditions, especially as the competition for the world's remaining oil gins up. Finally, there is probably not enough time to make and erect the number of wind turbines necessary to bring about any theoretical energy transition aimed at maintaining our current way of life. As energy blogger Gail Tverberg put it cogently, "it is appealing to have a 'solution' to what seems to be a predicament with no solution. In a way, wind and solar are like a high-cost placebo. If we give these to the economy, at least people will think we are treating the problem, and maybe our climate problem will get a little better."[10]

Solar electricity presents an additional set of daunting problems and complications on top of the intermittency issue it shares with wind (the sun doesn't shine at night just as the wind doesn't always blow). In fact, all the logistical issues that apply to wind turbines apply to solar: embedded energy in manufacturing the hardware, long mining and manufacturing supply lines, transportation of components to site, limited design life, exposure to the extremes of weather, technical expertise needed to deploy the hardware

---

10 Gail Tverberg, "The 'Wind and Solar Will Save Us' Delusion," *Our Finite World*, January 30, 2017, https://ourfiniteworld.com/2017/01/30/the-wind-and-solar-will-save-us -delusion/.

and the costs associated with all of that. So far, existing battery storage would not scale to accommodate the amount of solar electricity needed under any hypothetical "green energy transition" schemes. Solar has interesting geophysical limitations: for instance, in places assumed to be ideal for it, such as the Arabian desert or Arizona, the ground temperature actually gets too hot for solar panels to operate efficiently, while windblown dust frequently coats the cells' surface and requires continual human attention to keep performance up. Though solar panels don't have moving parts, they are subject to degradation from the very photovoltaic forces that make them work, plus rain, hail, temperature changes, and the jostling of winds. Bird and insect poop are even a problem. It's one thing to have an array on your own roof or out on your property where you can care for the panels personally. It's another matter when a company has to maintain a gigantic array of hundreds or thousands of panels. The materials get old and tired. Entropy never sleeps.

Solar panels require silver to operate. Silver is the best electricity conductor and has other physical properties that make it easy to work with. The proposed rapid growth of solar electric technology coincides with the peaking and decline of producible silver around the world, whether it is mined directly or is a by-product of copper, zinc, lead, and gold mining. Silver production already suffers from declining ore grades—you have to mine, crush, and process more and more rock to get the silver—and the green energy transition movement will find there simply isn't enough to build all the solar panels proposed in their scenarios.[11] One can also infer that the price of silver will become less affordable as the demand for solar panels increases. There are other types of solar cells in development that don't use silver, but at this point they are of negligible commercial value.

The bottom line is we will be able to set up a lot less wind and solar electric infrastructure than the public is being teased with these days. It will probably be done on a far more modest scale than the techno-optimists think—perhaps the village or community scale, at best. And we may only retain the ability to do even smaller-scaled high-tech wind and solar for a

11 Zolton Ban, "Not Enough Silver To Power The World Even If Solar Power Efficiency Were To Quadruple," *Seeking Alpha*, February 9, 2017, https://seekingalpha.com/article/4044219-enough-silver-power-world-even-solar-power-efficiency-quadruple?page=6.

limited period of history, say until the oil industry loses its viability. I stick by my hypothesis that solar and high-tech wind power as we know them today are wholly a subsidiary of cheap oil. The wind power of the future may be more like seventeenth-century Dutch windmills. The solar power of the future may take the form of hay grown to feed horses, oxen, and mules.

The "green" propaganda industry regularly ladles out breathless reports about "breakthroughs" in battery technology that turn out to be too good to be true. There have certainly been improvements in electric battery performance and price since 2005, but most dramatically for electric devices on the smaller end of the scale: laptop computers, mobile phones, Bluetooth speakers, even cars. At the larger "utility" scale, it's a different story. In 2009, General Motors paid to license the use of an "improved" battery made by a startup called Envia. But then the GM engineers actually working on an electric car reported that the performance fell far short of the touted innovations. The new battery also had the peculiar problem of mysteriously changing its voltage discharge, making it unusable. A year after inking the deal, GM management canceled it.[12] Elon Musk's Tesla car project had better results in their partnership with Panasonic, which developed a new generation of lithium-ion batteries that cut the cost by half while increasing storage capacity by 60 percent. However, the cost of Tesla cars still remains higher than the struggling middle class can afford, and there are serious problems with the stability of Tesla's batteries—they sometimes burst into flame.

Recently, both MIT and Stanford University announced promising new developments in liquid-sodium batteries, claiming the cost could be lowered by 80 percent with the same storage capacity as a lithium-ion battery. The improvements are generally made by making changes (sometimes miniscule) in the chemical composition of the anodes and cathodes, the negative and positive poles that give and receive electrons through the electrolyte. Other experimental formats include gold nanowire batteries that can supposedly be recharged 200,000 times without degradation; solid-state lithium-ion (replacing the explosion-prone liquid-lithium batteries now in commercial

---

12 Kevin Bullis, "Why We Don't Have Battery Breakthroughs," *MIT Technology Review*, February 10, 2015, https://www.technologyreview.com/s/534866/why-we-dont-have -battery-breakthroughs/.

use); graphene batteries (graphene is a patented two-dimensional carbon nanomaterial the thickness of an atom), which enjoyed peak hype in 2016 but have yet to show up in any commercial products; and lithium-oxygen batteries, which were first developed in the 1970s and have not reached the stage of practical application in half a century.

Other exotic formats are still in the labs. These include copper/plastic foam, ultrasonic, organic peptide units, aluminum/air, sand, carbon-ion, and a battery with a "microbial fuel-cell" with a human urine electrolyte, developed by grants from the Bill & Melinda Gates Foundation for third world people with extremely limited resources. There may be further breakthroughs in battery tech, but that doesn't mean they will evolve commercially—or soon enough to make a difference. It's also possible that we will see only incrementally small refinements in existing formats that do not add up to viable commercial applications.

Home solar installations do not necessarily come with battery storage. Many outfits are designed to just feed the electrical grid. But batteries for off-the-grid outfits, or for backup when the grid goes down, are generally not much different from the kind of battery you would have gotten in a 1925 Oldsmobile—a big, heavy, lead-acid box. Commercial wind, solar, and solar-thermal schemes can also employ geophysical storage.[13] For instance, some of the electricity generated during the day can be used to pump water up to a reservoir at a higher elevation, which can then be released at night to run turbines. (The process itself entails a loss of a percentage of that energy.) Air can be injected and compressed into geological vessels like salt caverns, and released later at pressure to run generators. The trouble is that these require special geologic conditions. Many places are flat (especially where the wind blows most steadily, like the Texas plains), and salt caves are not evenly distributed across the landscape.

Finally, there is the issue of the electric grid, a genuine wonder of the world. Americans rarely think twice when they enter their home and flick on the light switch. We just take it for granted that the juice is always available.

---

13 *Solar-thermal* refers to installations of mirrors or lenses that focus sunlight to create heat to drive electric generation. There are many experimental formats.

That is not the case in many other societies around the world. Intermittent or unreliable electric service would be a catastrophe for American business and a trauma for households. And it may be increasingly the outcome as we move forward.

The US grid is in poor shape. The American Society of Civil Engineers just gave the entire US energy infrastructure a grade of D+. This includes whole vast networks of power plants, transmission lines, substations, etc. One researcher estimates the current depreciated value of the grid at between $1.5 to $2 trillion, and its cost of replacement at a daunting $5 trillion.[14] Any way you look at it, the grid is nearing obsolescence. We're faced with a choice of rebuilding the system as it's currently configured or redesigning it in some new way on the assumption that "new technologies" will make that possible. These days, we run the electric grid on coal (38 percent), natural gas (27 percent), nuclear (19 percent), hydropower (7 percent), wind (7 percent), and solar (2 percent), with geothermal and biomass making up the rest. Note that this represents electric power only, apart from the fossil fuel products used for transportation, heating, heavy machinery, etc., that make up the rest of the US energy diet.

A popular proposed solution to the decrepitating grid is "distributed power." The basic idea is to generate power as close to the customers as possible—as opposed to the current system, which requires "dispatching" electricity long distances over high-voltage power lines in order to "balance loads." These days, more and more home solar units feed into the electric grid. This causes problems for the power utilities. The more solar households reduce their bill from the power company, the more the company has to charge the other non-solar-equipped households. The higher the bills, the more households are incentivized to install solar rigs and so on, a feedback loop. The power companies fight back with new, abstruse rate structures, or by reducing the rate they pay households for feeding the grid, and by charging connection fees for the privilege of selling the electricity your equipment

---

14 Joshua D. Rhodes, "The old, dirty, creaky US electric grid would cost $5 trillion to replace. Where should infrastructure spending go?" *The Conversation*, March 16, 2017, http://theconversation.com/the-old-dirty-creaky-us-electric-grid-would-cost-5-trillion-to-replace-where-should-infrastructure-spending-go-68290.

makes. In different regions of the country and depending on the time of the year (e.g., when air-conditioning demand is low), home solar units are apt to pump maximum electricity into the grid at times of the day when demand is lower, giving the power company surplus electricity that they have no means to store, but nonetheless have to pay for. Eventually, this disables the utilities' business model.

The upshot is that the current jerry-rigged system of partially renewable, partially centralized, and partially distributed electric power is unsustainable. In our grid system, two-thirds of the energy generated is simply dissipated as heat in the operation of the power plant itself, and another ten percent is lost along the long-distance transmission lines. More than two-thirds of those lines are thirty years old. The replacement cost is roughly a million dollars a mile. Now that we're talking about electrifying the Happy Motoring system, consider that each electric car put into service would be like adding a new house to the grid.[15]

Every few years, a new miracle is sighted on the horizon. All have proven to be mirages. When I wrote *The Long Emergency* around 2004, the big buzz was about a home fuel cell device the size of a refrigerator that would run on the most plentiful element in the universe, hydrogen, distilled from water or derived from natural gas. Every house would have its own electric power generation unit. It went nowhere. Nobody even talks about it anymore. There was a parallel idea back then over on the oil side of the energy story: a process called "thermal depolymerization" was going to take garbage and organic waste and distill it easily into high-grade crude oil. It went nowhere, too.

The world is becoming a more conflicted place. Old nuclear weapons treaties were dumped by the Trump administration, and old animosities are being stirred up—a revival of the Cold War with Russia, along with new currency and trade wars with China. The next big threat to our national life may be as simple as shutting down the electric grid. The expanding

---

15 Alice Friedemann, "Can the lights be kept on with distributed generation? 2015 U.S. House hearing on a reliable electric system," *Peak Energy & Resources, Climate Change, and the Preservation of Knowledge*, April 20, 2018, http://energyskeptic.com/2018/can -the-lights-be-kept-on-with-distributed-generation-2015-u-s-house-hearing-on-a -reliable-electric-system-2/.

"internet-of-things," in which so much of the equipment in our lives is tied into or controlled through the world wide web, poses interesting hazards. It's a straightforward case of Joseph Tainter's model for collapse: *the overinvestment in complexity with diminishing returns*. Computerized automation of the grid leaves it vulnerable to cyberattacks. Russia was suspected of causing widespread power failures in Ukraine in 2015, the year after a US-supported coup toppled the Russia-friendly president Viktor Yanukovych, and tempers were running high over Russia's annexation of Crimea.

A cyberattack on the grid could come from anywhere, even a lab of stoned graduate school pranksters or nonstate rogues halfway around the world. After a cyberattack—say, just messing up automated operating algorithms—the grid could come back online, even if it took a while. An electromagnetic pulse attack would be something else, very possibly a crisis that would go on and on, eventually leading to the deaths of large numbers of people. An EMP attack is an electromagnetic pulse that could be set off by a long-range missile with a nuclear warhead programmed to explode high up in the atmosphere, or a powerful bomb smuggled into a coastal port city on a container ship, or flown over the US border via Canada or Mexico in an ordinary airplane.

Here's one EMP scenario that's well within the realm of possibility. Depending on where the device was set off, many or few people might die in the initial blast. However, the electromagnetic pulse itself would fry all the electrical equipment in the region where it exploded. Everything from computer networks, power station equipment, municipal water and sewer systems, automobile and truck electrical systems, agricultural machines, airplanes and airports, dams, heating systems, hospitals, refrigerators, TVs, and radios—everything that depends on electrical components—would be rendered inoperable and unfixable. The chaos would be epic and accretive as the days went by: people would run out of food, drink unclean water, couldn't call their families, couldn't work or get paid, couldn't get medical treatment, and couldn't run any machines. A lot of the ruined equipment couldn't be easily replaced, including the generators and turbines for making electricity. We just don't manufacture much of that anymore.

Of course, an EMP attack on some region of the United States—the New York metropolitan area, the Pacific Northwest—would certainly be answered by a US counterattack. After all, we do have nuclear submarines armed with missiles and warheads circulating in every far-flung corner of the world. But now we're in the realm of nuclear war, World War III, and it's not necessary to spell out the consequences of that. Einstein put the knock-on effect succinctly: *World War Four will be fought with sticks and stones.*

Apart from a potential deliberate takedown, the fate of the US electric grid will follow the fate of our fossil fuel energy supply. As that becomes increasingly unreliable or economically nonviable, we simply won't be able to replace the parts of the grid that fail until, incrementally, it is no longer a working grid. When that happens, an awful lot of other things will no longer work either. And, one way or another, it implies a lot of attrition among human populations. That's the point where distributed power is the *only* option, meaning it will need to be created locally. The places that are able to cobble together hydroelectric generation from the leftover bits and pieces of the industrial age may be able to carry on something recognizable as civilized existence, at a much smaller scale.

In *The Long Emergency*, I stated that nuclear energy may be the only way to keep the lights on in the decades ahead—even granting the well-known hazards of atomic waste and various sorts of reactor failures from Three Mile Island to Chernobyl and Fukushima. But I'm no longer persuaded that we will be able to keep running our aging reactors, or build new ones, and I'll explain why. We pretty much missed the boat on that, in the same way that we missed the boat on building a high-speed passenger railroad system, or even refurbishing the old conventional railroad network already in place.

We are entering a period of impaired capital formation and capital scarcity. The money won't be there. The US has ninety-eight operable nuclear reactors, down from 104 in 2012. Nuclear accounts for 19 percent of the electricity we use. The US nuclear power infrastructure is aging out. The oldest operating reactor is Nine Mile Point 1 in upstate New York, which entered commercial service in December 1969. The average age in the US reactor "fleet" is thirty-seven years. Almost all of them are at or near the end of their functional design life. The newest reactor to enter service is Tennessee's

Watts Bar Unit 2, which began operation in June 2016. The next youngest operating reactor is Watts Bar Unit 1, also in Tennessee, which entered service in May 1996.[16] They are the exceptions.

There was a lot of controversy about the quality of the so-called Generation 2 reactors that represent most of the equipment developed during the heyday of nuclear energy: the 1960s and '70s. Among these were the General Electric Mark 1 design. In 1976, three GE scientists quit their jobs in protest over the design weaknesses of the Mark 1 containment vessel. The installations at Fukushima, Japan, were of this type. Many US reactors are the same type. Now, the Generation 3 reactor designs have come along with many improved safety and performance features. They produce less nuclear waste, feature more rugged containment, claim a longer design life of sixty years, and can be made much smaller. What's missing is the money to pay for them and the political will to prioritize a massive program to replace the old reactors and clean up the mess they left behind. Such a program could take a decade or more, even if we had the money and commitment. Keep in mind, as systems failure proceeds, things organized at the gigantic scale become more unmanageable, especially centralized government. The fact that little is being done now suggests that it is already too late—we should have begun this at the turn of the millennium. By the way, there have been no advances in deploying the oft-touted, supposedly cheaper and simpler thorium reactors, which I discussed in detail in my 2012 book, *Too Much Magic*. The old gag about atomic fusion (as opposed to *fission*) remains true: fusion is the energy of the future, *and always will be*.

Beyond the structural part of the picture is the social part. We should have every reason to expect that the contraction and collapse of our linked energy and economic systems will lead to considerable social disorder. As that occurs, it will be increasingly difficult to be confident about even managing existing nuclear energy. For one thing, the nuke plants absolutely depend on fossil fuels to service the reactors. Access to fossil fuel backup power generation is essential for running the reactors and for shutting them down

---

16 US Energy Information Agency (EIA), https://www.eia.gov/tools/faqs/faq.php?id
    =228&t=21.

safely. Outside power is required to keep the spent fuel rods from burning up in their cooling pools and spreading nuclear poison outside the facility. In a grid emergency today, nuke plants would have to run generators on diesel or natural gas to keep the reactors and the spent fuel stockpiles stable, and then what happens when the fuel for that runs out? For another thing, if social disorder means that workers aren't paid regularly, there is nothing to keep them at their tasks except a sense of duty—which may not be enough. At this point, if we were to take any action while we can, it should be to plan the orderly shutdown of our nuclear power plants so they don't create havoc when we have far less ability to deal with the problem.

POSTSCRIPT: DIGITAL NIGHTMARES

In my opinion, whatever upside there might be in Artificial Intelligence (AI) is far outweighed by the downside of it. AI is the epitome of lethal *overinvestments in complexity with diminishing returns.* Can we, please, not venture into trans-human monsterdom?

Anyway, there appears to be a race on between the development of really effective (and perhaps self-conscious) AI and economic collapse that will be sweeping enough to disable the attempts underway in its development. I believe economic collapse will win that race, that we will lose the ability to continue going down this path. It's the necessary corrective to a case of Faustian recklessness. The late Stephen Hawking was not too keen on AI and warned against its development strenuously in his last years. If I can be a little mystical about it, economic collapse may be Earth's way of defending itself against a kind of fatal insult to the planetary organism. Human beings have a variety of agendas and values, and human intelligence is dangerous enough; why would we want to develop the equivalent of an artificial life-form that is nothing like us, and which may embark on agendas of its own that we don't even understand? If AI was truly intelligent, as we understand intelligence, it would shut itself down.

The internet itself has become problematic enough, especially as it is increasingly commandeered for surveillance and control of the public. Big

Data has already become an odious racket, collecting information about everybody around the clock via their connection to the web and their activities on it—supposedly for commercial purposes, resulting in custom personal advertising on your browser screen based on analysis of your online behavior. So-called *consumers* (a diminishing breed, actually, as the middle class withers away) are being enticed by Big Data giants Amazon and Google to put "personal assistant" devices in the home, little cones or boxes that sit innocently on the coffee table. These nifty gadgets will play Beatles songs at your command and dim the lights in your bedroom automatically—zowie!—but who knows who might be listening to the conversation in your living room, or even watching what you do when these devices are integrated into home security cameras?

About the first thing the US government did with the maturing internet was construct the National Security Agency's facilities for collecting all the communications between US citizens—in other words, spying on us. They regularly deny that they do this, and their denials are laughable. China has an autocratic government that hardly has to answer to its citizens, and it has constructed a system explicitly and overtly for social control—and they're not hiding it from anybody. The so-called *social credit system* is already operating there. It will automatically tote up a "social score" based on each citizen's behavior, doling out layers of economic and social privilege to those who are most compliant in obeying the instructions of the government.

China endured a pre-computerized experiment in social control back in the 1960s called the Cultural Revolution. That was bad enough. The catch is that China's social stability depends on their ability to maintain the industrial economy that they ramped up in a little more than one human generation. And they have similar prospects for trouble with their fossil fuel energy supply as do the Western nations. They are on the same collapse path as everybody else. How will the social credit system work when China's economic tide goes out and the large, new middle class starts to suffer . . . and becomes politically restless?

The integration of social media, Facebook, Twitter, Instagram, and other apps with the cell phone phenomenon a decade ago seemed like playful, innocuous add-ons in the early going. The diminishing returns of these monsters have proven to be unforeseen havoc in American politics. Twitter and Facebook mobs now routinely destroy careers and reputations; erase the boundaries between facts, rumors, myths, and propaganda; and induce sociopolitical hysterias. The mercurial president uses Twitter for making the kind of policy announcements that used to be routed through layered filters of White House advisers, cabinet officers, and press secretaries. Now, any old thing that pops into his head enters the public consciousness directly via the president's thumbs, including suggestions of war with other nations. Social media, it turns out, amplifies and accelerates antisocial behavior among a population that was already having a hard enough time processing reality.

Facebook's founding president, Sean Parker, denounced the company's unanticipated social effects at the 2017 Axios conference, saying, "God knows what it's doing to our children's brains." He and his fellow social media innovators knew what they were building. "How do we consume as much of your time and conscious attention as possible?" he explained as their mission. "And that means that we need to sort of give you a little dopamine hit every once in a while, because someone liked or commented on a photo or a post or whatever . . . The inventors, creators—it's me, it's Mark [Zuckerberg], it's Kevin Systrom on Instagram, it's all of these people—understood this consciously. And we did it anyway."[17] A former Facebook vice president, Chamath Palihapitiya, speaking at the Stanford University Business School event that same year, said he felt "tremendous guilt" over his work on "tools that are ripping apart the social fabric of how society works."[18]

The cell phone itself has had ghastly transformative effects on interpersonal behavior. People literally experience a mediated existence through these screens that amounts to a simulacrum of being alive in a real place. It's not an enhancement of real life but increasingly a substitute for it. The cell

17 Mike Allen, "Sean Parker unloads on Facebook," *Axios*, November 9, 2017, https://www
    .axios.com/sean-parker-unloads-on-facebook-god-only-knows-what-its-doing-to-our
    -childrens-brains-1513306792-f855e7b4-4e99-4d60-8d51-2775559c2671.html.

18 Julia Carrie Wong, "Former Facebook Executive: Social Media Is Ripping Society Apart," *The Guardian*, December 12, 2017, https://www.theguardian.com/
    technology/2017/dec/11/facebook-former-executive-ripping-society-apart.

phone with its evolving functions is the epitome of mistaking the virtual for the authentic, one of the fundamental quandaries of the computer age.

At least these questions have stirred up debate among those in the thinking class. One important additional angle, however, is uniformly left out of the debate: how all this computer technology is really at the mercy of the fragile electric grid. All discussion about where digital technology is heading in the future ignores this sticky issue, and how it is tied to our fossil fuel supply and debt-based finance system. I would go as far as to say that the electric grid as we know it is surely going down, probably sooner rather than later, and even further to say we may be fortunate when that happens. The Faustian bargain was always understood to be a bad deal for the human race. We need a time-out from all this frantic digital innovation to reflect on what we think we've been doing.

I'll discuss the issues of climate change, planetary eco-damage, and politics in Part Three. Part Two of this book is a deep dive into the lives of a cast of characters who have corresponded with me over the years. They are people who are on board with the idea that we face difficult times ahead as the techno-industrial economy wobbles and crashes. They are all people who have decided to lead alternative lives and who are walking the walk of those lives. They have all bravely made hard choices, and their stories form the up-close and personal side of the long emergency whose consequences are at the heart of this book.

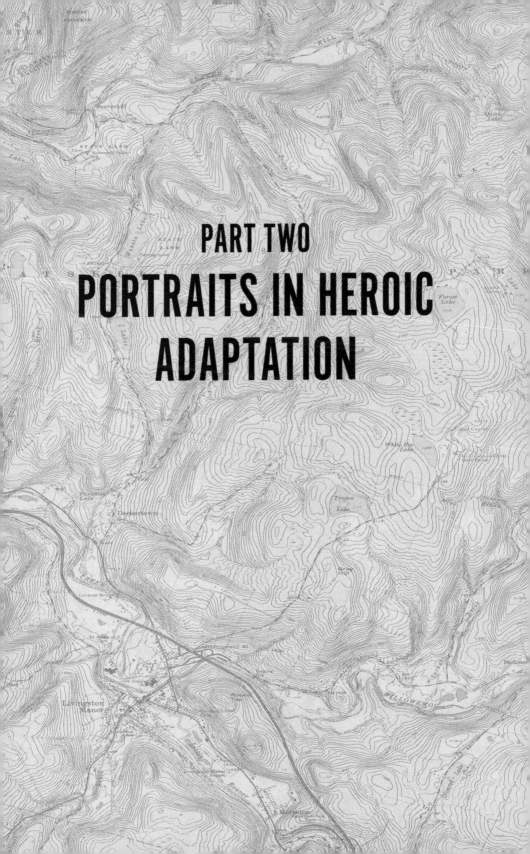

PART TWO
PORTRAITS IN HEROIC ADAPTATION

Chapter 3

THE GARDEN OF EDEN . . . APPROXIMATELY

Driving west about sixty miles out of Madison, Wisconsin, just past the town of Spring Green, where the occultish architect Frank Lloyd Wright built his communal studio-fortress, Taliesin, you enter the so-called Driftless Area of the upper Midwest. This is a region that was not covered by glaciers during the last ice age. And so it was not scraped pancake flat by the ice sheets like so much of the land beyond the Great Lakes.

The landscape looks superficially New Englandy, but the deeper you go into it, the stranger it gets. Ridges and plateaus yield to scores of intimate hidden valleys carved by streams that often vanish into a fractured geology of limestone caves and sinkholes. If they weren't forested, these valleys, or coulees as the locals call them, would look like the canyons and arroyos of the desert Southwest. Down in the valleys, cows graze the rugged hillsides like they do in Vermont. You notice that the barns and outbuildings look better kept than back east where I live, but Wisconsin was settled later, and the things in it are not as old. A lot of the houses, on the other hand, are mere dull suburban boxes of post–World War II vintage. Up high, on the plateaus and blunt ridges, it's mostly pure Midwestern agri-biz-style industrial crop farming, except for Mark Shepard's New Forest Farm, which most casual observers might not even recognize as any kind of farming operation they've ever seen before.

Mark Shepard calls himself a restoration agriculturalist or permaculturalist or agroforester or silvopasturist. There are many names for what he does because it comes out of several branches of a tree (pun not exactly intended) of principle and practice that started growing in reaction to the rise of the machine age in the nineteenth century. For three decades, he's been single-mindedly working to develop a method of getting food from the earth in a way that does not destroy it. He believes, as I do, that the current mode of techno-industrial society, including agri-biz, is heading toward some kind of collapse—at least, an inability to continue living the way we do now. He would like, as I would, to see the human project continue under a new disposition of activities and values, if only because there may be nothing like it anywhere else *out there* in the known universe, and even we humans may underestimate our own cosmic worth.

I met up with Mark on a hot mid-June morning at the produce-packing barn near the entrance to his 110-acre "research station," as he calls the place. He's a sturdy six footer and looks uncannily like the movie actor Bill Murray, with a cleanly cropped beard, as if Murray were cast in a biopic of Ernest Hemingway. The packing barn had been a hard-cider bottling operation for a few years. He'd planted or grafted 239 varieties of apples on the farm over the years. But recent laws had made it impossible for "little guys" to compete with the "big boys" for distribution—one of innumerable annoyances that handicap any enterprise below the giant, corporate scale. So he closed down the cider outfit and now focuses on myriad other activities including the crops he raises and sells, the tree nursery stock, public speaking gigs, writing a new book (a follow-up to his first, *Restoration Agriculture*, 2013), and paid consulting jobs with other people who aspire to do what he is doing.

The front of the barn was an office that looked more like a clubhouse, full of ratty old furniture. We set out at once to hike the property. The terrain of this Driftless Area is covered with a *biome*—a naturally occurring community of plants and animals—called oak savanna, a transitional ecology between the broadleaf eastern forests and the Great Plains. There's an overstory of trees, mainly in the family *Fagacae*, which includes oaks, beeches, chestnuts, and other flowering trees. The canopy covers 30 to 60 percent of the savanna biome. The rest is grass. A rich canopy of many layers or stories will promote abundant grass growth.

"Another overstory tree would be the *Prunuses*, the cherries," Mark explained. "These have shrubby forms in the lower story, which include plums. Hazelnut is one of the shrubs in the area. Crabapples and hawthorn are some of the medium-sized trees. The understory might consist of the cane fruits, raspberries, and blackberries. Gooseberries and currants are in the shade-tolerant understory. Grapes. All of this whole system, of course, is producing a lot of woody biomass. And so that gets decomposed by various different decomposers, which are fungi. And then there's obviously the grass and the prairie grasses, and who eats that? All the animals do. And so the savanna biomes around the world are the home of the majority of the world's mammals. There are fewer mammals in a forest than there are in a grassland. There are more animals in the savanna than either.

"And the human being basically loves the savanna," he continued. "That's our homeland. It's the Garden of Eden. It's where we came from, the Old-uvai Gorge, if you're into the whole out-of-Africa genesis thing. And if you want proof that the savanna form is appealing to human beings, look at the golf course, you know, the lawn with the trees. It's open underneath. We've got grass. We can see any predators come. We can climb trees for safety if we have to. There's food on all the trees and shrubs, bushes, and vines. And there's animals around. If we eat animals or use animal products, it's all there. It's a complete habitat for humans."

I followed Mark down a swerving alley planted up with chestnuts on one side and hazelnuts on the other. He estimates that he's planted 250,000 trees and shrubs on the property since he started the project in 1995. Viewed from above, New Forest Farm is a tapestry of these curving alleys of trees and shrubs. The pattern is based on a water-management system designed by the Australian Percival Alfred "PA" Yeomans (1905–1984) for landscapes with water-scarcity issues. He called it the Keyline System.

"It was difficult to read, dry," Mark said unironically. "The idea is to take whatever rainfall is coming out of the sky and you don't let it leave your sight. If it rains, if there's overflow running away down a drain or down a ditch or a gully, that's a wasted resource. Plants can live without almost every major nutrient out there—they can live with a deficiency. They can't live with a deficiency of water. They die. So any raindrop that falls on this farm, the goal is to capture it, and spread it out, soak it in, store it in some kind of ponds or tanks for use later when we need it."

This Driftless Area of Wisconsin gets about thirty-four inches of rainfall annually. The average for the USA as a whole is thirty-nine inches, but of course that includes extremes like the Great Basin of Nevada and the Florida Everglades. (It's forty inches annually in my corner of upstate New York.) Thirty-four inches might not seem problematic. But on Mark Shepard's place, this rainfall tended to come in a few quick, short, intense bursts that punctuated long dry spells, sometimes lasting all summer, when the plants needed rainfall most. And when the rain did come in a deluge, or violent storm, it would run off quickly in this landscape of plateaus and hollows, ultimately into the Mississippi River, thirty-odd miles to the west, taking topsoil and nutrients with it.

Mark described the Keyline System using the features of his hand:

"See, if you look between your fingers, the crack between your fingers, where it's that little fleshy web—that's, like, a key point in the landscape. So on almost every project where I do water management, I'll find the key points, mark those, and then we reference off of that when we're manipulating water on the site."

In actual practice it means sculpting the landscape with excavating equipment to create a pattern of swales (channels) and berms (raised ridges) with the alleys of trees and shrubs planted along them to capture as much water as possible. Between the alleys, Mark often plants lanes of annual crops: grains or nitrogen-fixing legumes, with a careful method of tillage that only minimally disturbs the soil. Such a rye crop, he says, can put twenty thousand pounds of carbon back into the soil in a season. The legumes put nitrogen back in. Every year, he lets some of these annual crop lanes go fallow. The soil of New Forest Farm has improved tremendously in twenty-three years. When he got the property, it was a pretty distressed combo of overfarmed cornfields and overgrazed pastures.

He was pretty dauntless about experimenting with any kind of tree or shrub that might produce food in this mode of permaculture. He planted 135 different perennial edibles including blueberries, apricots, pawpaws, and schisandra berry, a medicinal fruit native to northeastern China, reputed to retard the aging process and increase energy. He was also running trials with English walnuts, American butternuts (native to Wisconsin), and seven kinds of Korean pine-nut trees. His method was straightforward: "Mortality is important for finding out what doesn't die." Of all the forest food plants, the chestnut was a favorite.

The American chestnut (*Castanea dentata*) once dominated the eastern hardwood forests from Mississippi to Maine. It was particularly thick in the middle of that range, now comprising the states of Kentucky, Tennessee, Virginia, West Virginia, and Pennsylvania. Twenty-five percent of the trees in the Appalachian Mountains were American chestnut. The *mast* (the nuts that fell every season) produced a prolific amount of feed for the native animals. Early Americans used to brag that a squirrel could leap tree to tree through the chestnut canopy from Maine to Georgia without ever touching the ground. The wood was fabulously versatile for building anything from houses to furniture to musical instruments.

An Asian bark fungus that carried a killing blight was tragically introduced on some trees imported from Japan at the start of the previous century. It was first noticed on specimens in the New York Botanical Garden in the Bronx in 1904, where professional horticulturists kept a keen eye on the plantings. After that, the spores traveled about fifty miles a year via the air, eventually affecting 220 million acres of American forest. By the mid-twentieth century it had killed three to four billion trees. A few blight-resistant specimens still grew here and there. Oddly, a large grove of about 2,500 trees planted by early settlers in West Salem, Wisconsin, forty miles northwest of New Forest Farm, survived long after all the eastern chestnuts had died off—the disease finally showed up there in 1987. Other small stands still exist in Michigan and Oregon, and occasionally individual surviving trees are found here and there. Blight-resistant hybrids are sold by many nurseries, including Mark Shepard's.

While western Wisconsin is at the margin of the American chestnut's range, by all appearances Mark's are growing nicely. The blight usually shows up in the first couple of years, and many of his trees planted early on are tall, thriving, and bearing nuts. His mission, he says, was to *imitate* the North American savanna, not to reproduce the original version exactly. He began planting New Forest Farm with American chestnuts hybridized with Asian trees, which have a lot more resistance to the blight but grow smaller and tend not to be as hardy as the American natives. Once the trees are planted, he said, his input costs are zero: no herbicides, pesticides, or chemical fertilizers. Even pruning is minimal. His method for selecting out the most hardy, productive, and successful trees is "complete and total neglect—the survivors win." Same with the hazelnuts. He considers the chestnut an excellent food crop and claims that many of the products made from corn these days could be made from chestnut flour. Corn is a majestically resilient and versatile food crop, but it is hell on a biome. And the way that the agri-biz boys grow it is washing away the topsoil of the American breadbasket at epic rates. It can't go on. And, as we already know, things that can't go on, don't.

We hiked uphill slightly, past a 1951 International Super M model tractor parked among nettles, black raspberries, and elderberries. It had been reposing there a couple of years. Mark said he could get it running again, no problem, with a few hours work. We followed another alley with its berm and swale down into a little gulch where some hot electric-wire fencing was strung into a compound for the eight sleek Tamworth hogs he was raising for the year, a durable, hardy breed originally out of England, well suited to forest grazing. A little farther down the lane, the smart and curious hogs found us. They were adolescents, still young enough to be cute. In season, the hogs would eat a percentage of the nuts and fruits that dropped to the ground, but they also got grain fed twice a day in June, before the drop. Mark had named them all after firearm manufacturers: Winchester, Remington, Ruger, Browning, et cetera. A ways farther down we came to their headquarters in a shady little spot, a wooden shed shelter with a crude feeding trough. Hogs are tough on their furnishings. They like to bust everything up, whether it's edible or not.

A hundred yards farther we came to the house that Mark built, where he actually lives, and went inside for some water. It was a pleasant wooden structure, with an improvised feel, like the hippie houses of years gone by, raw and unpainted, but put together with love, care, and skill. He had a solar electric unit rigged up for it—the house was totally off the grid otherwise—and the lights and other household equipment ran off direct current (DC), though he planned to install an inverter that would allow him to run normal alternating current (AC) appliances. The downstairs was an open plan with the kitchen dominating. It had a bachelor pad look, and when I asked where Mark's wife, Jenn, was, he kind of choked up and admitted they'd broken up recently after about thirty years of marriage and two sons, now grown and off in the world leading their own lives. He really didn't want to say more about the situation, though you will learn their history soon.

VIROQUA

Later, we hopped in Mark's beat-up old Subaru wagon and drove over to the town of Viroqua, sixteen miles west, for a fancy dinner. The Midwest is full of dying small towns. The town actually nearest to New Forest Farm, Viola (pop. now 650; 825 at its height in 1940), has got little commercial infrastructure left besides a couple of gin mills and a post office. What suburban sprawl didn't kill off, big-box shopping did. Ironically, now that the town lacks the most rudimentary shopping, cars are what allow homeowners to remain living there, driving to distant jobs for income and to Walmart for supplies. At least until more recently, when plain, old, bad weather kicked in.

A tornado churned through the town in 2005, the same week that Hurricane Katrina swamped New Orleans, diverting virtually all FEMA help down there. Viola, which is located on the bank of the Kickapoo River, then suffered two devastating floods in 2008 and 2016. Many ground floors and storefronts were flooded in the first one, then renovated, then flooded again, and not renovated—a couple of hundred-year floods in eight years. One abandoned storefront there had a mysterious sign in the dusty window that said "Epitaph News." We drove on through.

Viroqua (pop. 4,395) is another story. By today's standards the town is doing quite well. It's the seat of Vernon County. That helps (government jobs). It has a celebrated farm-to-table restaurant downtown, the Driftless Café, and a couple of other gourmet-type eateries, the Rooted Spoon and the Tangled Hickory Wine Bar. It has a Waldorf School. It has an Orvis store (there are lot of trout streams in the Driftless Area, which attract anglers from Madison), and an immense secondhand bookstore in a great big masonry tobacco-drying building—harking back to the days when leaf wrappers for cigars were grown and processed there.

Mainly, what Viroqua has is the Organic Valley cooperative, a gigantic national organic food distribution enterprise that started up in 1988 when Midwestern farmers were suffering from a lethal wave of foreclosures due, in large part, to the policies of Reagan-era secretary of agriculture Earl Butz, who declared to America's farmers, *Get big or get out!*

The Driftless Area was hit hard, in part because the rugged topography lent itself more to small farms than giant ones, and many farmers just couldn't get big. For years, Vernon County was the third-poorest in the state. Lately, its population is growing again. Organic Valley has two thousand grower-members, $1 billion in annual sales, and 932 employees, many of whom live in Viroqua with good salaries and benefits. Organic Valley is the nation's largest farmer-owned organic cooperative and one of the world's largest organic food brands. It brought little Viroqua back to life. Mark Shepard has been a board member since the 1990s.

We settled into a corner booth at the Driftless Café. Luke Zahm, the owner, had trained at the Culinary Institute of America in New York's Hudson Valley, and he'd worked at some of the better restaurants in Madison, the state capital and seat of the main campus of the state university, before coming back to his hometown to open the bistro. He'd even been nominated for a prestigious James Beard Foundation award. The dinner menu was a kind of coded organic mystery, with items like "Mushroom Mike Chicken of the Woods," "Ocooch Mountain Elderbelly Gastrique," and a cheese called "Blue Mont Bandaged Cheddar." A couple of beers went down fast at the end of the very hot day and opened up the conversation to the long, strange trip that brought Mark Shepard to the Driftless Area.

He was born back east in Lancaster, Massachusetts, a rural township north of Worcester and about forty-five miles west of Boston. His father, Byron, was the son of a Maine hunting guide.

"My grandma was the lodge matron, which meant she did everything: sewing, cooking, cleaning, managing the place," Mark said. He had a storyteller's innate sense of dramatic velocity. "And then the Depression kind of slowed business down. World War II killed off the rest of the sports, and the guy who owned the lodge sold it, and that brought the whole family down to Massachusetts where they could work in a factory. My grandma actually worked in a sweatshop making pocketbooks for years and years. Imagine sewing pocketbooks every day, starting on the treadle sewing machine, and then more and more advanced sewing machines through time. And then my dad's father died of lung cancer I guess from the tanning chemicals in the leather mills where he worked tanning leather—kind of tanned his lungs from the inside out."

Mark's mother, Agnes, was a Vermont farm girl. Her family lost the place during the Depression and they, too, moved to Massachusetts to work in the factories. His father, Byron, eventually became a machinist and toolmaker.

"He would make the assemblies to make a particular part, like the wheels on a Tonka truck. You got these plastic wheels, these plastic hubs, these metal axles. And so you make all the different chutes and conveyor belts and bins and hoppers and levers and Rube Goldberg–type inventions. He made the machine that put those wheels together."

His father was intensely interested in gardening from his own boyhood around the Maine hunting lodge. The lodge owner was an associate of Rudolf Steiner's circle of anthroposophists, a humanistic/spiritual quasi-religion of ethics, education, and character development. One offshoot of anthroposophy was the Waldorf School network. Another was the organic "biodynamic" gardening movement. Years later, in Massachusetts, out square-dancing one night, Mark's parents met the anthroposophists John and Helen Philbrick, authors of the book *Organic Gardening for Health and Nutrition* (1971), and Byron started to go to meetings of biodynamic gardeners down in Duxbury, Massachusetts, often taking young Mark with him. By then, he was gardening in earnest according to biodynamic principles.

The Philbricks were connected to a larger social circle that included the legendary Helen and Scott Nearing, the authors of many books on homesteading and gardening. Scott Nearing was a political radical maverick so cantankerous that even the communists and socialists kicked him out of their confabs. In the 1940s, with the help of a million dollars of inheritances, the Nearings carved out a handmade homestead in Vermont—they specialized in handmade fieldstone and concrete houses, barns, and walled gardens—and eventually resettled in Maine. In 1954 they published *Living the Good Life: How to Live Simply and Sanely in a Troubled World*, a book that eventually became a sort of bible for the 1960s back-to-the-land movement. The biodynamic circle included other leading figures in the anthroposophical movement: Heinz Grotzke, Hartmut von Jeetze, and Herbert H. Koepf, all gardener-authors. The Nearings would frequently come down from Maine to the monthly meetings. Mark remembered them vividly, especially the weather-beaten guru Scott Nearing, who would eventually live to eighteen days shy of his hundredth birthday:

"I think this was in Chicopee, Mass," he said. "I was a ten-year-old kid. I wanted a soda pop. And it was an embarrassment because we were the natural foods associates meeting with all these organic growers. Helen and Scott are there. And the Philbricks are there. What are people going to think if Byron Shepard's son drinks a soda pop? He finally relented. We go out to the Coke machine. And there's these grumpy old German guys being mean to this really nice little old man [Scott Nearing]. I was fascinated by Scott. I'd heard him talk before and I thought he was brilliant. Really. You know, amazingly well-spoken. And it's, like, let's go get my Coke, and my dad goes, 'Sshh Sshh Sshh.' You know, like, stop! Evidently, I stood there while Scott and these old German anthroposophist biodynamic folks debated something. I don't know what it was. And they were heated and angry, and he was really civil and polite and handled everything that they threw at him."

And eventually young Mark got permission from this gang to drink a Coke.

Otherwise, he says, he lived a normal suburban childhood. When the time came, he entered the Worcester Polytechnic Institute on a scholarship, supposedly for a degree in mechanical engineering (the kind of work his dad did). He was profoundly unhappy doing that.

"I failed to thrive because I didn't enjoy it. Achhh, it was, like, sterile."

He dropped out and got a job at a company, Natick Labs, that made equipment for the army, testing the new Kevlar infantry helmets before they were introduced to the troops.

"I was the test monkey. I was the guy that ran the product through the final test phases. The whole project had nothing to do with me, my skills, or interests, or talents. It was, like, I just walked into it, and there it was."

Around that time, he got a bee in his bonnet about going up to Alaska. He'd read an article in *Yankee* magazine about the federal government closing down the Homestead Act in Alaska forever.

"Not the agricultural homesteads where you'd get one hundred and sixty acres, but the home site, which is five acres, [or] a business headquarters site and trade manufacturing sites for eighty acres," he explained. "And all you have to do is stake out the land, file a fee to say that I actually staked out this land, describe where it is. Then you go build a dwelling and dwell in it for five months out of the year for three years. And so I read that and I'm kind of like thinking to myself: *I'm twentysomething. If I don't do this now I'll miss the opportunity, because it's closing in a year. I've got to do it. And I don't want to be like these guys at work who are in their fifties and sixties and have heart problems and knee problems and back problems, high blood pressure. I want to live outside in this beautiful natural place.* And so after pondering it for a considerable amount of time, I sent away for all the material. And found out that the least remote parcels were three hundred miles from town and five miles off the nearest road thirty-five hundred feet up the side of the mountain. It's like, well, that's not really practical. So I kind of mothballed it for a while."

Working at the Natick Labs, he'd met a girl. Jennifer, or Jenn, was a student at the University of New Hampshire and worked at the lab over Christmas and summer vacations. Mark asked her for a date and she said she was "taken." She'd come and go and he would ask her out again. She'd say, "Not yet."

"It's like, who answers 'not yet'?" he said, remembering his exasperation. "It's either yes or no, isn't it?"

Not long after that, he quit Natick Labs in a huff.

"I slammed my clipboard down on the desk. I said two words, the second of which was 'it.' I stood up and everybody else looks around the office and they go, 'Oooo, what got into Shepard today?' and I walked off the job. My boss said, 'Okay, give me two weeks' notice.' I said, 'No I quit. I'm outta here.' And he says, 'Mark my words, you're making the biggest mistake of your life.'

And, Lenny Flores, I'll grant you your point, from your perspective. Yes, I made the biggest mistake in my life. But I think that's the mistake that made all the difference for me."

We knocked back the last of our beers, paid the bill for a fine meal at the Driftless Café, and made to leave.

"So then, after I quit, it was when I was driving home—because I was still living with my parents at the time—and thinking to myself: *Maybe I have made a terrible mistake. I've got no job. I've got no education, really, that's worth anything. How do I save face here?* So I quickly enroll in Unity College in Maine to study ecology, because I wanted to be outdoors."

He was eleven credits shy of the bachelor's degree when he became obsessed again with Alaska. There was still time to go up there and stake a claim.

FROM THE DEEP NORTH TO THE MIDWEST

The next day, even hotter than the previous one, Mark and I drove from New Forest Farm to the town of Cashton, fifteen miles north of Viroqua. Mark had a load of asparagus that he had harvested in the early morning to drop off at Organic Valley's new distribution center in that town. On return, we took a different route on the back roads and soon began to see horse-drawn hay wagons and other signs of Amish life along the way. The region is 50 percent Amish now, Mark said. They've been fruitful and multiplied. They started buying land there in the late 1960s when farms were cheap, and now there are more than three hundred Amish families in and around Vernon County, one of the largest Amish settlements in the country.

Of course, a snapshot of Amish farm life reveals many differences between their ways and the conventional agri-biz scene, but none so moving as the sight of many hands working at something together. This used to be one of the great consolations of rural life in pre-agri-biz times. Humans are deeply social, and sharing hard work with others makes it bearable at worst and fun at best. The crews of young people riding in the horse-drawn hay wagons in their distinctive "plain" garb looked both happy and healthful—especially in contrast to the stunningly obese, unhappy-looking customers we saw at the convenience store back in Cashton, where we'd stopped for coffee a few minutes before.

Nowadays it's gotten to the point where one lone agri-biz farmer can spend a day mindlessly sitting in the air-conditioned cab of a giant, million-dollar harvester guided by GPS, watching movies on an iPad while the machine does all the work, including the mental labor of deciding when to turn the rig down a new row. It's barely necessary for a human to be mentally present while the job gets done. The giant harvesters also come with giant mortgages. The biggest "input" in this mode of farming is borrowed money, a.k.a. debt.

Farmers, by and large, used to be proud of their mechanical abilities to fix their own equipment. But the giant new machines can't be fixed on the fly. They require elaborate diagnostic equipment when something goes wrong. The diminishing returns of technology suggest a high potential for ennui in that scheme of things. In 2018, the Centers for Disease Control and Prevention ranked national suicide rates by profession, and farmers have the highest rates by more than 30 percent. These days in America, in many professional domains, we are far removed from the experience of working physically with others, and the emotional toll is substantial. Circumstances in the decades ahead are very likely to change this, and we're fortunate that the Amish are keeping these practices alive for others to emulate.

We followed the course of the Kickapoo River much of the way back. Some years ago, the federal government came up with an "economic development" scheme using the Army Corps of Engineers to dam up the river, in order to create a "recreational area" and encourage a real estate bonanza of waterfront lots for vacation houses on the reservoir. But the proposal aroused a storm of protest among everyone from farmers to hippies to Indian tribes, and the case was hung up in court for years until the government decided to just forget about it and dropped the project. In the meantime, Mark said, a lot of the designated area returned to the wild. Eventually part of it became a state park. The river outfitters moved in. And it did become a recreational area of sorts, but without the dam. And without disrupting the Amish farming that had developed all around it in the meantime.

When we got back to New Forest Farm, the temperature was above ninety and we settled back into the kitchen at the house, in the deep shade of the trees Mark had planted around it, and, over some cool drinks, he told me about how he finally got up to Alaska. At the end of that spring semester at Unity College in Maine, with the window closing on the Homestead Act very much on his mind, he caught a ride to Colorado with his roommate:

"And so we got to Denver—because he's going home for the summer—and I got a little part-time job for a while. And I was kind of looking at finances, and thinking about my level of courage, and I was scared shirtless—can I say without the *r*?"

Mark generally avoids profanity.

"I had enough money either to get to the Pacific Northwest and see some big trees or I could get down to the Grand Canyon area. I opted for the Canyonlands and I hitchhiked off in that direction. And through a series of miracles, I bumped into a guy named Bill who had just quit his job as a mechanical engineer working for the US military. I'd worked at a US military lab, and I'm heading up to Alaska to claim some homestead land. And he was on his way up to Alaska in a van, and he had two sets of skis, two bikes, two sets of snowshoes, two sets of this, two sets of that, he's looking for a hitchhiker to go all the way up to Alaska with him. It was like: *I was supposed to meet you!*"

Mark and Bill journeyed into Alaska in the van to the claim area about three hundred miles northeast of Anchorage on the edge of the Wrangell-St. Elias National Park, as far as they could get in an automobile.

"You know I'm pretty good with a map and a compass. There's this river in our way. How are we going to get across this river? Well, all of a sudden, these two guys come across in a boat, a canoe. Wild and woolly, hair all over the place. Dirty, stinky, you know, rubber boots up to their knees. Shotguns slung over their shoulders. We struck a deal. They got to go to the post office, get some supplies at the local roadhouse. There's, like, a general everything store. It's got a garage, a liquor store, bar, a café for simple meals, grocery, showers, ammo, of course. Our fee for them, to guide us in, take us across the river, was a bottle of rum."

Once across the river it was another three miles in, some of it muskeg, the typical swamp or bog of the northern boreal forests, which you simply had to slog through in high rubber boots at that time of year. All of the valley land was already claimed, so Mark staked his claim on a high "bench," as the shoulders of the mountains are called. Bill absolutely hated Alaska because of the mosquitoes, which for some mysterious reason he had not anticipated. When they slogged back out of the wild, he did an about-face and hightailed it back to his home state, New Mexico, "where the trees are properly spaced apart," as Mark says he put it.

Mark filed the paperwork for his claim with the feds in Anchorage. Then, with his engineering and science background, he was able to get a job as an electrician's apprentice in the North Slope oil fields.

"Anyway, that was the offer on the table, but I'm thinking to myself, *I'm going to go back and see if my girlfriend wants to marry me.*"

He ditched the oil-field job idea, put a plane ticket on a credit card, and flew back east. It had been a component of his Alaska dream all along to have a female partner on board for the adventure. There was a catch. The girlfriend he was thinking about marrying was a woman he'd hooked up with at Unity College the year before. Then something weirdly fateful intervened.

"As I was walking into my parents' house, the phone was ringing and it was the gal from Natick Labs who I was really sweet on and who'd said she was 'taken' with her boyfriend, or whatever. We started talking and we kept talking. We were married within the year and moved up to Alaska."

This was Jenn.

"How did she happen to be calling the moment you were walking in the house?" I asked.

"Things happen," Mark said.

"That's bizarre," I said.

"Serendipity," Mark said.

"Synchronicity," I suggested.

"There are interesting things that happen in this world, in this universe, and for some reason that we don't understand," Mark said. "I'm okay with that."

It took another year of futzing around back east, with Mark working a data entry job at the Internal Revenue Service, to build up some cash reserves to return to his faraway land claim.

"We went to Alaska on our honeymoon and stayed for eight years and lived in a little tiny cabin in the bush."

Actually, it ended up being two cabins.

"We built one on my claim. And then [Jenn] ended up claiming land as well. I think she's the second-to-the-last homesteader in American history according to the 1860 Homestead Act. We would live in both of them depending on whose turn it was to do your 'settlement time.' And so we'd do five months in my cabin, five months in her cabin. But we had to earn some money in the summertime. In the fall I would go out early to my cabin, get all stocked up for winter, cut wood, haul supplies in, while she's still working in town. So for like a couple of months, I'd be out there alone. Then in the wintertime, the middle of winter, we'd shift, go down to her cabin, and I would leave early in the spring, and so she would be out there for two months in the springtime alone."

I remarked that it was pretty extraordinary for a woman in her early twenties to be willing to live that way.

"Yeah, she's an amazingly competent human being. I mean for a suburban gal from Wellesley, Massachusetts, this was about as far away as the moon. Well, I grew up gardening and the whole homesteading thing—Helen and Scott Nearing—and this is an ideal of mine. I knew Grandpa was a hunting and fishing guide. My mom's side of the family were all farmers living on the land. So I'm going to do this back-to-the-land thing. To me it was what I always wanted to do. Well, for her this was like an adventure way beyond adventure. Think about it: we're twenty-some-odd-year-old kids getting great exercise; breathing some of the freshest, cleanest air; the purest water you could possibly get. We're in outstanding physical condition. And then psychologically, it's beautiful. It's *beautiful!* And there were neighbors, and she would visit."

The nearest was a half mile down the valley.

Getting to the "general everything store" in the tiny hamlet of Slana, or occasionally "to town" (Anchorage, three hundred miles away), and back, was a big deal.

"There was a thirty-five-hundred-foot climb every time you came and went. On the way out, of course, you're coming down so you're walking faster.

Maybe a couple hours, maybe three hours at the most. Coming in it could take a long time, sometimes it could take half a day, because you're coming in from the road, you've got a pack, or in the wintertime you're pulling a sled behind you, or new snow has fallen in the wintertime, and you got snowshoes. And going up in the wintertime, if new snow has fallen, you're just slogging through thigh-deep snow up the mountain."

"How did you get through that muskeg when it wasn't winter?"

"You jump from tussock to tussock and try not to fall over—and you fall over several times. And it's mosquitoes. And it's just a mess."

I asked Mark what his twenty-odd-year-old self was thinking about in terms of what exactly they might do on their claim to make a buck, aside from temporary summer jobs.

"I was going to do a tourist service business."

"Like your grandfather at the Maine hunting lodge?"

"Yeah, right. And we actually had quite a few people through the years that would come and stay on a tent platform, and we'd take them up to the top of the mountains and turn them loose, and then they're on their own."

In the early going, Mark and Jenn packed in a lot of food from Costco and fished for salmon. But their relationship with the food system of the modern world preyed more and more on Mark's mind.

"That's what led me to an exploration of this whole idea of restoration agriculture. How do I live here? We're an outpost! We are an *export* outpost! The industrial machine is growing all these foodstuffs. They're putting it on trains and trucks and cars, then ultimately on a ferry that comes all the way up the Gulf of Alaska to Anchorage, then goes into more trucks that go to a store. My food is coming from thousands of miles away and it's a critically fragile system. We knew that that was the case because one time a storm had come up and it washed a whole bunch of containers off a barge. This was a supply barge going up to Anchorage. And so all of a sudden for the next month or two there's shortages of all kinds of stuff in all of Alaska. It really affected us in the bush because we didn't have a lot of money. We were only working a couple of months of the summertime at the most, and we were critically dependent on this food. So, when the stuff is not available and the price spikes, we were critically threatened."

Hunting up in the Alaskan wilderness was problematic.

"I hunted," Mark said. "But not for big game. We were able to get subsistence hunting licenses because we lived in the bush. It's only twenty-five cents. [For big game] you had to go to the area *where they tell you to go*. I can go shoot caribou, but I've got to go to unit ninety-two, which is like a hundred miles away from here. And we can't afford gas hardly to go to town to get toilet paper. Caribou, in the wintertime, are all clumped up as a herd. The herd was right at my house, going through our valley. Right there! But it was not unit ninety-two. If I shot a caribou in my unit, that's poaching. And it's real easy to follow blood trails, and you've got game wardens flying around in airplanes. They basically say that if you ever get caught poaching in Alaska, shoot the game warden and maybe you'll get away with murder. Hunting is a big financial deal for the state of Alaska. And because of the harshness of the climate, the wildlife populations are critical. You can really do a lot of damage with poaching and so that's taken seriously up there.

"So I tried to figure out: How can I live in this environment and feed myself from it? Obviously, I'm not going to be able to do it hunting and gathering because it's such a fragile, slow ecosystem. I'm going to have to, like, design this and really make it hyperactive with all these different perennial things that are growing. I definitely didn't want to do it the way my neighbor is doing it, which is the way that all agricultural civilization has done it. He started clearing the land. You cut the trees down, and you burn them up in the wintertime to have a little bit of open ground. Then you plow the ground and you grow a couple of carrots. You don't even get enough food to feed your family. So you destroy the ecosystem in order to grow some food.

"The sod up there was a mix of blueberries, raspberries, rose hips, Labrador tea. And so I'd have to rip up the sod food in order to expose the soil, get it blown away in the wind, washed away in the rain, and then throw a couple of hard seeds in the ground, hope that the birds don't eat the seeds, or the red squirrels, and then grow some little pittance of grain, which isn't that nourishing—and what? Subsist on bread by the sweat of my brow? Screw that! Let's figure out how to work with this environment. So I started experimenting with the different plant community types I've already mentioned.

Well, in a forest there are multiple layers: tall trees, medium trees, shrubs. There weren't any vines where we were. There was ground cover. There were currants, no gooseberries. And so why don't I modify the canopy as it is right now, go extra heavy planting food plants that are there. I started experimenting with a lot of very different pine nuts from different cold climates. And all of a sudden, it's like, wow, I can just imitate the natural plant community types, go heavy with the food plants, and harvest forever in a perennial system, so I can actually maintain this ecosystem, and not destroy the ecosystem in order to grow my food."

It worked to a point, but they needed more land than their claims provided to really raise enough indigenous foods. After eight years of struggle, they needed a new plan.

"We wanted to reproduce, and both Jenn and I had decided that we would rather our children be exposed to something more than living in the bush, chainsaws, snowmobiles, and shotguns."

Their first son, Erik, was born in Alaska. A short while later, they decided to move back east, to northeastern Maine, the region where Mark's grandfather lived and guided.

"My dad's family came across in 1820. There's no more people related to me up there that I know of. But I'd gone up there on vacation as a kid. So I was really familiar with it already and there were clear-cuts available for fairly affordable amounts of money."

In 1991, Mark went to Maine, bought property, and constructed an A-frame cabin for his family. They left a lot of stuff up in Alaska but brought a bunch of cherished tools back east.

"I've got a basic set that I can survive with wherever there's any kind of woody crop growing. I've got my grandpa's bow saw or bucksaw, a crosscut saw, two-man saw. One is a pit saw for making boards. Sharpening tools. A froe for making shingles. So I can go out anywhere and with these 1800s tools, just sharpen them up a little bit, and have firewood, building materials, all of that."

The new life was a bit easier in Maine: "That was pretty special because instead of being six hours from town and three miles off the road, thirty-five hundred feet up the side of the mountain, we were only an hour and a half from town, which is kind of nice. If you plowed your driveway regularly, you could get there all winter."

Mark Shepard was closing in on thirty years of age with a young son and another baby on the way, and he was increasingly dissatisfied with life in the wilderness, either in Alaska or Maine. He was still dogged by a wish to practice some kind of sustainable agriculture dissociated from the oil-and-debt economy. In the fall of 1994, he signed up for courses at Jerome Osentowski's Central Rocky Mountain Permaculture Institute, located at 7,200 feet altitude in Basalt, Colorado. Jerome had come to Colorado as a ski instructor in 1969. Ten years later, a health scare over hypoglycemia got him interested in healthy, fresh foods. Over the years he developed a compound of gardens and greenhouses where he was producing figs, papayas, and bananas in a climate so harsh that the bighorn sheep struggled for existence on the mountain shrubs.

Mark took the courses and got certified in permaculture design, but he found himself developing a view more radical than his teachers'.

"They'd sit down at a table and eat rice and beans for their meal. And I'd say, 'Hey wait a minute. I thought we're supposed to be permaculturists here. What are we doing eating annual crops, these annual grains and legumes.' 'Oh, but it's organic.' Yeah, it's organic, but they still have to destroy an ecosystem to plow up the ground, to expose the soil, to put the seeds in, to get these grains and legumes. And you still have the soil erosion, loss of wildlife habitat, no more carbon sequestration from a woody perennial cover, et cetera. And so I think it was the section on agroforestry; at the end of my session I had the floor and I said, 'All right, this is what we've got to do. We have to design perennial food systems that are ecologically designed, that mimic nature in its structure. But it also has to be effective so we can actually plant it at scale, we can manage it at scale, maintain it, we can harvest it at scale. By at scale, I mean at scale that's large enough to actually feed people for real—no joke, no hallucination. And that means we can do it on hundreds of millions of acres. So look at our farmland right now. Over 50 percent of the United States of America is agriculture right now. So we need to be able to do permaculture at 50 percent of North America. And feed three hundred million people in this country. That requires a lot of stuff and you can't do it with little berry bushes in your backyard. We're talking hundreds of millions of square miles of food everywhere, food ecosystems on every piece of terrestrial planet.'"

After Mark made his spiel about *what we have to do*, two fellow students, Roland and Rand, came forward and volunteered to join him in just such a venture, even to help finance it.

"They're like, 'Wow, we need to do this! We've *got to* do this! I've got some money. I don't like these investments [stocks, etc.]. Let's throw it all together, buy a piece of property!'"

They set up a limited partnership. None of them knew exactly where to set up the proposed operation.

"And this one guy was from Milwaukee, he said, 'I'll go to southwestern Wisconsin. It's really beautiful down there and the real estate is affordable.' 'All right. Why don't you go check it out.'"

Rand put a down payment on what would become New Forest Farm. Mark ponied up his share. Roland dropped out and the other two split his share. Mark and Jenn moved to the Driftless Area of Wisconsin. They settled in a hippie commune called Dreamtime Village in West Lima, six miles down the road from the farm property. I asked him how he came to join a commune after so many years of autonomy with a nuclear family in the wilderness.

"I was a Nearing-ite since I was a little kid, and we were at least associated with the whole biodynamic anthroposophy school movement, which is very collaborative, very community-oriented. You need people. We're not islands unto ourselves. We're very social beings. Even if I spent a lot of time out here alone, I go to town and hang out with people. When we were in Alaska, we joined with several other settlers and formed what was called the Mountain Front Association in order to deal with things like the Bureau of Land Management, access rights, trail maintenance, stuff like that. Then when we moved here, we actually were in an intentional community. The idea was: we're going to use this property [the farm] as the food source for the village. I had all the tools to make it happen. And the situation was that it was a very unmotivated group of people. They didn't want to come out and actually work and create something. They wanted to hang out with stoners most of the time."

In the beginning, Mark worked hard not only on the farm project but on improvements to the buildings of the commune.

"I've got all the tools to fix things. I'm going to start working on their buildings and fix them up. So now they'll be habitable. I made a ferro-cement bathroom that you can just hose out because you've got twenty people using the bathroom plus guests."

One building was an old school.

"Listen, I don't like to diss people. I don't. They had a fire truck in the gymnasium of the school building. One of their buildings caught on fire and they called the Lafarge Fire Department. They say, that's not our jurisdiction. That's Viola. They called Viola. Oh, that's not our jurisdiction, that's Richland Center. Oh, that's not our jurisdiction. It burned to the ground. Now did you hear what I said at the beginning of this? They had a fire truck. They couldn't be bothered to use it. They couldn't be bothered to hook a freakin' hose up to a fire truck and turn the thing on and put out a fire."

Meanwhile, Rand, the other partner, had a swift change of plans when his wife inherited property from her grandfather in Assisi, Italy. Rand went over to check it out and didn't return to the USA for eighteen years. Mark bought him out and also began quarreling with other commune people over the share they were paying to live at Dreamtime Village, considering all the construction he was doing. In 1995, he had enough and moved the family to the farm. They lived in a camper while Mark began to build a proper house. They spent one winter living in the decked-over foundation and then finished the house. Finally, he set about transforming the scrubby old hayfields, cornfields, and overgrazed pastures into an edible landscape mimicking the oak savanna biome that had been there originally. It was land that had been maintained for millennia by fire, bison, and the native inhabitants for whom the "tension zone" ecology between forest and prairie was a stable and sustainable place to live.

THE PERMACULTURALIST

He'd been at it on New Forest Farm for a quarter century when I came to see him. Music had been a big part of the social scene there over the years.

Mark played guitar in a series of rock-and-roll and bluegrass bands. The boys were out of the house, on their own now. Jenn had departed recently, too, and was living, Mark said, over in Viroqua, though he hadn't seen much of her for a while. He hadn't gotten over the split, and there wasn't a new lady in his life. He'd become something of a guru himself on the alt-farming scene after publishing his first book about his methods, *Restoration Agriculture*, in 2013. He was about to publish a second book, *Water for Any Farm*—a conscious tribute and an update of P.A. Yeomans' classic *Water for Every Farm*, adapting the keyline water system originally developed for Australia to American conditions. He remained active in the still-expanding Organic Valley Cooperative. But he was spending a lot of time on the road giving talks and doing consulting jobs, designing farms with people who aspired to do his kind of permaculture.

The evening before I left to return to upstate New York, there was a social gathering on a property adjacent to New Forest Farm, where a couple hosted a croquet game and weenie roast every Friday night in season for decades. They weren't farmers. Rather, they were a mom, dad, and son construction outfit that specialized in taking down old barns and outbuildings and reassembling them on new sites. It was a few days before the summer solstice, the longest day of the year. The lowering sun glowed reddish through a humid haze that tempered the day's fierce heat. By twilight it couldn't have been lovelier, with long shadows on the fields and pastures all around, and swallows winging for bugs. There was good food and plenty of beer. A fire pit was stoked and crackling. The twenty or so people on hand were a mix of Boomers and Gen Xers, with more in common culturally than you might imagine, especially when it came to recreational activities. They'd rigged a floodlight on a wire over the nicely manicured croquet court so the game could continue well into darkness. I had hardly played the game since college. The Driftless folks made short work of me. I headed back to my B and B in Richland Center when there were just a few salmon-colored streaks left in the western sky.

Just before we went down to the croquet scene, I asked Mark my last question:

Now what . . . ? I explained that I was not so much after his personal plans for the years ahead, but his view of where the life of our time was headed, especially given the converging problems of peak affordable energy, climate change, population overshoot, and the gathering disorders of American politics.

"It's a good question. I can't not do this. I mean, everywhere I go I'm going to be doing this. I'm going to be planting ecosystems, food ecosystems, at scale, that will actually feed you, not just a little demonstration project in your backyard. And I think that the future for agriculture and humanity, whether we're [looking at] a high-tech future, or . . . a low-tech future, agriculture has to and will become more ecologically oriented. How this planet works is [that] if you are not ecologically in harmony, you get selected against. Something happens if you mess with an ecosystem. You tear it apart, tear it apart, tear it apart. Pretty soon there's a collapse. If we don't go toward a perennial ecosystems-based agriculture, whether we're high-tech or low-tech, we're going to collapse, because there's no way to avoid that. People have been pretty resourceful for the past few near-extinctions of human beings. We've somehow pulled out of it. But there's so many people right now and our impact on the planet is so significant and so huge that the hiccups might be unexpectedly bigger than ever. And if we don't have the natural biological integrity from which our species came, whether we were evolved, adapted, created, or whatever, it doesn't matter necessarily where we came from, when we came, how we got here—we came from a whole, complete, intact perennial ecosystem that included disturbance in annual crops, and old-growth redwoods four thousand, five thousand years old. If we don't go to a whole, complete ecological integrity of this entire biosphere, it's game over. Maybe there'll be some people around. Maybe not. I don't know. That's not for me to determine.

"I'm here to start pointing the way. We've got to not *go back* to the garden or *go back* to nature. We have to *go forward* with nature. With all of our brothers and sisters and plants and animals as companions along the way because they were all here for some reason that we may not understand. Everybody's got a role to play in this. We need the birds, we need the frogs, we need the weasels. Tell me, somebody please tell me why we need ticks. But they're here. There's something going on. Ticks are kind of important to somebody."

"How about the deer fly?" I asked, referring to a notoriously obnoxious denizen of the eastern forests.

"I can handle deer flies," Mark said. "You know how you deal with a deer fly? You grab an oak twig and just stick it behind your ear and they won't bite you."

Now I know.

Chapter 4

A NATION OF ONE IN A PLACE CALLED LIMBO

He called himself "Thwack" in the comment section of my twice-a-week blog, *Clusterfuck Nation*. Of course, I must review the comments and police them a bit, since cranks and pests come aboard sometimes. There are a few regulars with extreme views whom I allow to post because they are, at least, polite in presenting their unappetizing ideas, and squelching them would keep a certain political reality that we must contend with out of the arena where, at least, they are exposed to light and argument. A couple of these are white nationalists.

Thwack showed up a few years ago and began jousting with these characters. He caught my attention because he was witty and intelligent, and he did a great job of tweaking them while presenting his own ideas about the world in a thoughtful and original way. For instance:

We need a common "American" culture based NOT on what you look like, but how you act (behavior).

There should be constant discussion and dialogue to determine what this common culture is, and what it should be. Diversity should not mean you get to do whatever the fuck you want . . . ?

If I put on a Tee-shirt with the letters "USA" and an American flag, and I go walk the planet . . . what should people think when they see me coming?[19]

It was only after a while that Thwack revealed himself to be an African American. As such, he had a good deal of sport with the various "right thinkers" and "bad thinkers" on the blog. For example:

Did you know before Morgan Freeman became everybody's favorite Magic Negro, president, and the voice of God himself, he was a straight up gangsta killin black savage who slapped Hos and threatened Superman himself?
Cover your eyes white people
This is gonna hurt

He linked to a YouTube clip from a 1970s Christopher Reeve *Superman* movie, featuring young Morgan Freeman as a pimp threatening one of his girls with a broken Yoo-hoo bottle.

I have access to email addresses of these blog commenters "backstage" on my WordPress platform. So, I started corresponding with Thwack. We also had several phone conversations. Eventually I learned his name, though he asked me to use a pseudonym, Josh Wickett, for the purposes of this book, because he didn't want to become a target for those who might be offended by his views. I also learned a few concrete details about him. He lived in the Baltimore ghetto. He'd been making a living around Washington, DC, for many years doing carpentry, home renovations, landscaping, and painting houses, and the financial fiasco of 2008 put him out of business for a few years. He ended up in nearby Baltimore where a property owner he knew offered him a place to live in exchange for work, fixing up a bunch of row houses in the Johnston Square neighborhood, close by Green Mount Cemetery. He was mainly interested in making music and sent me audio files he recorded. Many of them were covers of classic sixties and seventies rock and roll. (He's a Dylan fan.) The guy could play the electric guitar. We got to know

---

19 James Howard Kunstler, "Not So Happy Motoring," *Clusterfuck Nation*, March 22, 2018, https://kunstler.com/clusterfuck-nation/not-happy-motoring/.

each other. When I determined that my next book would include in-depth profiles of interesting people leading alt lifestyles, I wanted to include him in it and he agreed. So, on a morning in early May, I flew down to Baltimore.

I'd been in Baltimore several times in recent years. The beleaguered old city still had some charm in the close-in neighborhoods like Fells Point, Canton, Federal Hill, and up Howard Street, where quite a bit of renovation had gone on, but the city had more recently become notorious for its large and violent black ghetto. Millions of people have watched *The Wire* on HBO, the cinema verité series about street drugs and political corruption set in the Baltimore slums. And then there was the 2015 death of Freddie Gray in the back of a police paddy wagon, followed by riots, arson, and looting. The city had become a national poster boy for urban dysfunction.

I met Josh in the lobby of my hotel in the early afternoon. He was lean and lithe, dressed in shorts and a short-sleeved shirt with sandals and a straw hat. His hair hung down in dreadlocks. He was fifty-three years old. He saw me sitting on a sofa and smiled broadly, both of us apparently amused to recognize that we were real people in a real world, not just internet figments. We ducked into an adjoining Starbucks for coffee, then went up to my suite and hunkered down at the square table in the sitting room with the digital recorder. I asked him to start at the beginning. He said his eyes were bothering him from the tree pollen in the air this time of year and apologized for sniffling. His voice was soft and his accent was straight-on standard American, except when he used some black locutions for effect, and his speech was free from casual profanity.

He was born at Fort Benning, Georgia. His father, an Alabama farm boy, joined the army, went through officer training school, and eventually made it to the rank of lieutenant colonel. His mother was from Houston, where her father worked as a stevedore on the ship channel. Her family was Catholic. His parents had met at college, Tuskegee University, formerly the Tuskegee Institute, near Selma, Alabama. Upon its founding in 1881, Booker T. Washington was brought on as principal, aged twenty-five. Washington dedicated his life to black self-reliance, laboring against the rising tide of segregationist Jim Crow laws that saturated the former Confederate States into the twentieth century. He was convinced that if former slaves developed skills, entered professions, and demonstrated competence, they would eventually achieve standing and respect from the dominant culture—even in the face of Jim Crow. Tuskegee graduates tended to be self-starters, and perseverance in adversity was one of Washington's prime credos. Through the twentieth century, the political pressure to correct racial injustice mounted. After the Second World War especially, what with adopting its role as "leader of the free world," the US had to confront the enormous moral failure of a segregated society. The Civil Rights era ensued, and the legal basis of Jim Crow was dismantled, immediately followed by Lyndon Johnson's "War on Poverty," with its unanticipated and destructive knock-on effects of welfare dependency and single parenthood. Josh's parents were far enough along in life to avoid those traps.

"They lived the American Dream," Josh said of them. Before he was born, his father was stationed on an army base near Hamburg. "My mom had a maid in Germany. I'm using that as an example of the whole American Dream thing, you know, to go from her mom working in other people's houses."

His mother was an only child, but she had eight children—five girls and three boys—Josh being the middle boy and close to the youngest of all the children. His parents had passed away in the early 2000s, and by the time we met in Baltimore, he had not spoken to any of his siblings in seventeen years. But he remembered his childhood as happy and normal.

"In the military, you move every three, four years," he said. His memories begin around the early 1970s. The family moved to Dale City, Virginia, when his father was posted to the Pentagon.

"That was the whole middle-class suburban experience. A new subdivision with families and children, very, very typical. There was a little bit of social engineering going on. This was the start of that open classroom routine where you don't have separate classrooms. You can see the other classes and it was also year-round schooling. You go three weeks and you're off for three weeks. So, all through the year you're going to school. I had a very good childhood and a rather magical one. It was that good. Just no worries, having brothers and sisters, and having fun, and playing in the woods, and building forts, and go-karts, and everyone's doing well. It was mostly white because it was a new subdivision. And they were expanding it. When we moved there, there were a couple of sections already done and we were like the third section. But they were making a fourth, fifth, and sixth, and so we got to play on construction sites, you know, have dirt-clod fights. I had a great time."

After that, the family moved to Fort Jackson, outside Columbia, South Carolina.

"The thing I remember about the military life is that everyone had a father. It's very structured and there's no anonymity."

The US Army had been desegregated by Harry Truman in 1948, but in the 1970s, most of the officers living in the officers' housing at Fort Jackson were white. Most of his friends were white kids. I asked if he was self-conscious about that.

"No, no. Growing up, if you look up to your father and he has a good position, you don't really think about it that way. Because in the military you could see other soldiers your dad didn't know that would have to salute him. You see that and you realize that it's built on a meritocracy. You go by the rank of the person and so I understand why children, especially male children, may be damaged by not having their father. I never had to think about that. But I'm sure it affected me, having a father who I could look up to. Even though we had our differences, that foundation is you're not insecure about your masculinity."

His father was a public information officer. Earlier in his career, before Josh came along, he had done two tours of duty in Vietnam, and he spoke Vietnamese.

"I have no idea what he did there, but as an officer it was a dangerous place. My dad had the advantage of going into the military after World War II, where they were probably looking to promote and advance black soldiers. So, I think that was some of it. And he was competent and good at his job."

He describes his parents as "conservative."

"I figured out later that his conservatism was probably not so much based on the military, but an older version of the split between conservative and liberal black people. It has to do with W. E. B. Dubois and Booker T. Washington."[20]

---

20 W. E. B. Dubois, a Massachusetts native and the first African American granted a Harvard PhD, came to oppose Booker T. Washington's program for accommodation with segregation and pushed instead for social activism and equal rights under the law.

With eight children in the household, his mother was a stay-at-home mom.

"She was a military wife. You basically support your colonel, no matter whatever disagreements you have. You don't air that stuff in public. And, you know, Catholic—so there's a conservatism to that, too. You never heard them battling. Part of that is based on military culture where if you've got to argue with your wife, how are you going to command troops? You don't argue with your wife. You don't tell somebody something twice."

The colonel quoted General Douglas MacArthur on "duty, honor, country." He had a framed copy of the speech on the wall. He put a flag up every day at home and took it down at night. Josh's family stayed at Fort Jackson for four years, his middle school years. He remembers the Carolina pines, school trips to Fort Sumter where they gave the kids some "Confederate" money. His school was off the base, so he got to mingle with kids outside the military orbit. I asked if he brushed up against racism there.

"Not really. That was school. And so when you're in school it's like everyone kind of has the same Adidas tennis shoes and the dungarees and stuff. But I did notice some definite Southern culture. That was my first experience with people who knew the history of the Civil War."

Meanwhile, all the girls in the family were going off to college. But none of the boys would. In fact, the boys would not finish high school. Something happened.

"This is where things kind of broke down in my family having to do with the male children, and no one wants to talk about this. I kind of don't go into it much myself. I've never really gotten to talk about it. You know, in some families things are uncomfortable. You don't get to dialogue about questions like this unless you pay a therapist thousands of dollars and even they may not know. I've been on my own personal journey of discovery of figuring things out. It's almost as if that trajectory of upward mobility turned backwards. It's like, you know, some of us do better—except there's the sexual component where none of me and my brothers graduated from high school or went to college. It's not that we're stupid. We could have. But that's just the circumstance. I didn't drop out or get expelled. I wasn't bad. Let's say I left. There was no support for continuing on. You can kind of tell when your parents are motivated to support you in doing something."

Mainly what happened was that his father stopped getting promotions. They left the friendly confines of Fort Jackson and moved to Houston.

"The whole thing with my situation, with dropping out of high school, had to do with the fact that in the military, once you get to a certain rank, if you don't go to the next level, you have to retire. I think that my dad's retirement had a lot to do with it, because suddenly—and this probably happens to a lot of men who have been in the same career for thirty years or something—they lose all motivation for doing anything. I'd say that my situation would have been different if he had made the next rank up, which is full colonel, because then you're still in the game and you've got to do all these things. And it's one of the dangerous things about adopting the culture of your job too much. If you ever leave then you may have this hole and not know what to do. The military is a total head trip because you walk around in a uniform, guys you don't know are saluting you. And then you're a civilian. You walk down the street. Nobody knows who you are. I mean you can dress up and everything but nothing beats having your name out in front of your house. I just felt like the whole structure of the family just kind of . . . everyone was kind of confused. Because that's what happens when you have this hierarchical structure. If something happens to the head, you know, everyone's lost. And it's not like, in the military, your mother is not going to step in front of your dad to do stuff.

"Something changed for the worse and it was just like a blue funk over the house. I can't talk to people in my family. It's painful. So, I'm talking to you. I went to eleventh grade and kind of just went to school every day and did nothing. Played guitar or didn't show up that day. The school in Texas had a shooting range. I took ROTC (Reserve Officers Training Course) in tenth grade. And what else did they have? The smoking circle. If you had permission from your parents—and they didn't really check—you could smoke cigarettes. They had an area that was right near the school and you had to be within that circle to smoke. And I befriended some of the people that hung out there because it's one of those places that would accept anybody."

Music was becoming an important retreat for him, guitar especially. He'd been through a few other instruments.

"As part of that whole striver's culture that I was telling you about, my parents in fifth grade told me the next year you have to play an instrument. You have to. I was like, really, I get to play an instrument? Yeah, you pick an instrument. I picked flute. That got vetoed. I like to whistle and I can do most of the birds around me, so I'm like: I'll play the flute. [Low voice] 'No you're not!' And I didn't realize at that time that, you know, what are you? A faggot? I'm in sixth grade. I don't know what a fag is. My dad didn't say that but now I think back on it . . . And he was kind of right because all the flutes were girls."

As it happened, his father played guitar and a little piano. There was a guitar in the house, a Fender Mustang that the colonel brought home from Vietnam, along with a Fender Princeton amp.

"If I had it now, it'd be worth a few thousand dollars. So he would play, and this is what annoys me when some people say rock and roll is like white music and country music is like, you know, cracker music. His style of playing was to put that tremolo on and then just play one chord, strum one chord, and sing the verses and you strum another. It's very country. And guess what: he grew up on a farm and back in those days you listened to what was around you, like Charley Pride. People think country music is like a white thing, but in the South, you know, country is *country*."

He says his mother was "traumatized" by the changes induced by the colonel's retirement, that the marriage and the household suffered. Josh is of the view that his father's career might have been derailed by what today would be called a sexual harassment charge. He also believes that his father internalized his rage at the system— which he'd been so loyal to—turning on him, and that he took it out on his sons, turning his disappointment back on them and failing to support their own progress in the world. To what extent Josh internalized his father's personal tragedy I can't say, but it had obviously affected his view of how to find a place in the world—since the most seemingly secure place of all, the US Army, had proved to be so grievously unfaithful to the family. I infer that because otherwise wouldn't a troubled adolescent like Josh, coming from a military family, be urged to join up, if only to gain some maturity and bank some GI bill educational futurity? But his father, the colonel, did not steer him toward that option during these anxious years when Josh felt his family imploding and lost interest in getting on with a formal education. But he did find his way to personal autonomy, and he retained a sense of discipline from his upbringing that eventually allowed him to put a life together on his own terms.

He stopped going to high school altogether and, faced with the choice of just "being tolerated" in the household, or crashing at friends' houses around Houston, or becoming an autonomous adult in a world that was less comfortable than the structured life of the military, Josh decided to bug out of Houston altogether. One of his sisters invited him to come stay in New York, to discover if there might be some opportunities for him there. He accepted the offer. She had some connections with one of the actors' unions that enabled him to get a job at the now-legendary restaurant Windows on the World, which occupied the top several floors of the North Tower of the World Trade Center. The waiters there were almost all actors, good-looking young people.

"I was a page, and what you do is you stand in front of the elevator. When the elevator comes up, you direct the guest to whether they're going to the restaurant, or the hors d'oeuvrery, or the ballrooms, if there's a wedding or something. You get a little suit with jackets that have the double buttons like a bellhop. Black shoes and black pants. I did that for a while and that was interesting. After that I worked a variety of different jobs, just kind of hustling around. One was doing inventory for how retail stores close. They have people going in there and counting everything—kind of BS jobs like that. And then the ex-wife of my sister's husband worked at a studio, a [television] commercial studio, and they needed someone to do the janitorial things while the janitor went on vacation. I offered to do it and it paid well. I ended up staying on there doing other things, eventually becoming the messenger because the messenger retired. So that was very interesting to see that a commercial that didn't last thirty seconds takes a few days to make."

He did that for three years. Eventually, he moved out of his sister's loft downtown and moved in with workplace roommates at different places around the city, way uptown in Washington Heights and then over in

Brooklyn where he stayed at the gloomily named Transient Hotel with tiny rooms and a bath down the hall.

"I've always thought that that would be a cool name for a band," he said.

Our brains were both getting a little fried after a couple of hours recording. Josh was also getting nervous about how the meter was doing where he'd parked his car up the block. We went to check on it and have a look around town.

OUT AND ABOUT IN THE HOOD

His car was a cherry-red 1995 Chrysler LeBaron convertible with a janky passenger-side door that required some coaxing to open. The meter had expired, but there was no ticket under the wiper blade. We drove the back-streets up to Josh's neighborhood. It was fun cruising around with the top down, though the ancient LeBaron seemed to handle with all the grace of a ferryboat. He took me over to see the new Open Works community center built recently in a bygone industrial district. Approaching it, Josh had some rather unkind things to say about a gang of preteen kids we passed hanging out nearby with their bicycles. He said that when he'd gone to Open Works previously, he'd caught the same kids trying to vandalize his car. They glowered at him riding with me, but Josh didn't say anything to them directly, in keeping with his personal code, which we will get to shortly.

The Open Works center had been built around 2015–16 with support from the Robert W. Deutsch Foundation. Its mission is to "create safe, affordable, and accessible space for Baltimore's creatives." The operation also styles its patrons as "makers," in an attempt to promote a return to economic activities with a productive purpose, geared toward personal development, especially craft skills and manual arts. It was, curiously, very much in the spirit of Booker T. Washington's original Tuskegee Institute.

On the outside, the building was an architecturally uninteresting box jazzed up with giant painted graphics. But the interior was appealing and impressive. The main lobby was a bright and expansive performance space with a café at one end, a raised stage at the other, and a lot of moveable furniture. Everything was sparkling and new. Down a hallway there were well-equipped studios for wood, metal, and digital fabrication, 3-D printing, textile arts with sewing machine stations, painting and sculpture studios, and a computer lab. The center charged $125 a month for members to use all the facilities and tools for up to eighty hours a week, and lesser monthly fees for just using one type of maker studio.

It seemed to me a very good idea after decades of economic-development fakery and welfare dependency. Baltimore had been a mighty industrial city through the nineteenth century and into the mid-twentieth. Like so many other places in the nation, it let all that slip away. Or perhaps it is fairer to say that something had just come to an end. That *something* was industrial mass production at the giant scale. First, it moved across the border, then across the ocean. Now, there is every reason to believe that we may be leaving the industrial age behind, along with the cheap energy that drove it for two hundred years. A common assumption in the media has us moving into a dazzling robotic nanotech future powered by human "innovation"—plus some mystery "renewable" energy source as yet undiscovered. But what if that is not so? What if we are moving into a period of energy and resource scarcity, capital scarcity, population contraction, and relative hardship—the conditions that I described in my *World Made By Hand* series of novels? Wouldn't it make sense to teach young people manual skills, to train them as artisans? And by that I don't mean *artisans* in the fey sense of boutique consumerism; I mean preparation for a world where it is crucial to know how to make things after mass production fails, a world without social safety nets.

By late afternoon, we drove over to Josh's place. It was a lone row house on a desolate street in which most of the other row houses were long gone, with nobody around, as if a war had happened and then a plague had passed through. The house appeared to be a wreck from the outside. The inside was not a whole lot better, and we had to struggle over some rubble and wreckage to a battered stairway to get to his quarters upstairs. The appearance of squalor was necessary camouflage to keep looters and burglars out, he said, giving the impression that there was nothing of value inside.

It turned out, Josh had a lot of stuff: guitars, amps, computers, power tools and hand tools, electronic equipment, plenty of things that a crackhead or junkie might convert into ready cash. This was the place where he recorded his music videos. There were also workout stations, barbells, a boxer's heavy bag hanging from a naked joist, a full-length mirror. It had the look of an extraordinarily well-equipped prison cell. Indeed, Josh had commented earlier that he viewed black life in America as essentially "prison culture." I asked if he'd ever been in jail. He said that he'd never even been inside a police car. But he viewed racism in America as being absolutely structural and pervasive, and he based his comportment on the need to survive in a fundamentally hostile culture.

"I dump on black people being fuckups, but racism is a real phenomenon. And you have to come up with strategies to counter it because you can't just ignore it."

Some years ago, he'd chanced upon two figures, quite obscure to white America, who molded his view of his personal predicament and led him to develop his own code of conduct. The first was Frances Cress Welsing (1935–2016), daughter and granddaughter of black doctors (her mother was a schoolteacher in Chicago) and herself a Howard University–trained physician, who worked professionally as a psychiatrist in Washington, DC. In 1970, a lively time in black politics, she published an essay that got a lot of attention titled "The Cress Theory of Color-Confrontation and Racism (White Supremacy)." Years later, she included it in a book of essays called *The Isis Papers* (1992), which posits racism as a global campaign of white supremacy against all other races and peoples. She believed American whites especially were waging a chemical and biological war of "white genetic survival" on blacks, citing crack cocaine and AIDS as two of the then most recent atrocities, as well as the feminization of black men in order to thwart black reproduction. She held that no appeals to morals or reason would work to avail the defeat of white supremacy.

Welsing, in turn, had been highly influenced by Neely Fuller, Jr. (b. 1929), a Korean War veteran who worked as a security guard in the US Bureau of Engraving and Printing in Washington. Welsing and Fuller met at a Washington black power event in the seventies. Fuller was the author of a book titled *The United Independent Compensatory Code/System/Concepts: a textbook/workbook for thought, speech, and/or action for victims of racism (white supremacy)*.

Fuller, at eighty-nine, is still active on the scene, writing and speaking as I am putting this chapter together. He has a set of clear foundational principles supporting his prescriptive advice to black Americans, starting with the assertion that there is only one form of racism in "the known universe," and that it is "white supremacy." He regards relations between blacks and whites as "a total disaster" because "[t]he people who have the ability to eliminate racism do not have the will to do so, and the people who have the will to do so do not have the ability."[21] His conclusions were succinct as he explained recently on talk radio.

> Divide everything into one of two categories: constructive, nonconstructive. And try to make everything that you do, in every area of activity—economics, education, entertainment, labor, law, politics, religion, sex, and, if someone is fighting you, war, someone opposing you, someone saying something harsh to you to your detriment, deliberately and unjustly—try to always, before you do or say anything, think about what the result will be, because it's only one of two categories: constructive or nonconstructive. Either something will work for you or against you. This will apply to every person, every creature on the planet, but particularly when it comes to people of color.[22]

Josh had adopted Fuller's code of conduct for its commonsense value. It was, at bottom, a conservative code in the sense that it insisted on the discipline and regulation of each individual's personal agency—the actions, beliefs, and thoughts that they themselves controlled, without having to coerce other people. Hence, it was quite the opposite of identity politics now, in which coercion of others is paramount, the drive to tell everybody else how to think and what to do.

---

21 Produce Justice, https://producejustice.com/bio/.

22 "Neely Fuller and Mr Bobby Talk About Intelligent Resistance," *The Compensatory Concept*, August 8, 2018, hr 1, http://www.talktainmentradio.com/shows/archivedpodcasts/compensatoryconceptpodcasts.html.

"I don't like using the word 'conservative' but I don't have a better word to use," Josh said. "A lot of my advice to black people is first, before you start doing anything actively, start cutting things out that you think are not helping."

These were some of Neely Fuller, Jr.'s basic codes, which Josh Wickett had made his own.

- Don't argue with a black person, not even on the internet.
- If you're going to be near another black person, have a reason.
- If you want to do something, pick out a definite time frame to do it and try to stick in that time frame.
- Be consistent.
- Don't lie.
- Stop "name-calling." (Fuller says: Among black people, it oftentimes leads very quickly to fighting and killing. The result of name-calling is never worth the grief that such practices promote. Instead of calling a person a "liar," repeat what it was that the person said, and say that what was said was not true. Then explain why.)

Some of the rest were simple injunctions to stop: cursing, stealing, gossiping, idling, snitching, fighting, and killing. It suggested a kind of behavioral jujitsu, a way of avoiding maximum injury with minimal effort, mainly by opting out of the psychological games embedded in a dysfunctional culture—including all the mutually destructive interactions between whites and people of color.

Some of Fuller's prescriptions went beyond shorthand, and were both detailed and surprising, given the shibboleths of our time. For instance, some of his political ideas:

- Do not depend on, or ask others, to give you respect. Do not demand that others respect you. If you want respect, give it to yourself.
- If you are asked a question and you do not know the answer to that question, always say you don't know.

- Avoid attending any meetings, assemblies, etc., of any kind at which white people are not welcome.
- Buy land if you can, and sell land only to buy other land.
- Use your education only for the correct purpose. Use all that you can learn in such a manner as to best promote justice and correctness.
- Avoid being sidetracked into arguments about "capitalism," "communism," "fascism," and "socialism." Focus your attention on racism. Talk about racism in all of its aspects and all of its effects.

Josh Wickett had found the answers to many of life's perplexities in Fuller's concept of what is often called *structural racism*, just as Fuller had developed his theories and codes about racism over a long life in the twentieth century—including the many structural changes that took place along the way, such as the Civil Rights campaign, which produced major changes in law and social relations, though perhaps disappointing outcomes in economic relations. I had also lived through this period and had developed my own theories about race relations in America, which I will lay out more fully in Part Three of this book.

What I could not and would not do was argue with Josh Wickett about his credo or the ways he had arrived at it. He was obviously intelligent and, at fifty-three years of age, his life was rich in experience, too, and it was deeply different from my experience. What I liked about Neely Fuller's code was that it emphasized behavior based on the adult executive function of the human personality—the ability to size up persons and situations and rely on one's personal agency to optimize decisions about the project of leading a rewarding life as a sentient being. I am certainly aware of theories about an inborn sense of racial and group identity, and the group allegiance or aversion it implies, and how that might be the locus of what we call "racism." Which is to say that it is not a figment of imagination. But I do not necessarily see it as the overarching condition of human life. And, of course, everybody would benefit from being treated fairly, justly, rationally, and politely. All civilized societies possess some version of the Golden Rule.

What I thought Fuller overlooked, or discounted, was the reality that life is difficult for everybody, of all races, and presents problems that are hard for everybody to overcome, not just people of color. Josh's childhood in the military milieu suggested to me that he had a deep emotional attachment to clear codes of conduct and was greatly relieved to discover Fuller's ideas later in life, after suffering the breakdown of his family and his estrangement from that military setting, and its codes, that had been so comprehensible and comforting.

"It's almost as if a black parent cannot tell the truth about the full depth and breadth of the racist system without discrediting themselves," Josh said, "because when you tell your son or daughter that you can't protect them—you can't tell them that! You've just got to look the other way and hope that they get through somehow."

THE ROCKY ROAD TO BALTIMORE

It was getting on supper time, and we were both hungry. When we left Josh's house, thick gray smoke billowed up into the sky a couple of blocks south, another house burning down, he said, a not-uncommon spectacle in the hood. We got back into the red LeBaron and drove uptown, around the sprawling Johns Hopkins campus, to West Thirty-Sixth Street in the artsy Hampden neighborhood, a street locally known as "the Avenue" for its many bistros and boutiques. We got an outdoor table at the Avenue Kitchen & Bar, a hipster restaurant serving up some of the renowned bounty of the Chesapeake Bay, and picked up the thread of his life's journey at the point that he left New York for Washington, DC.

Josh felt "aimless and rootless" in New York in the mid-1980s, so he moved to Vienna, Virginia, a part of the country where he had fond memories from childhood and where he thought he might feel more comfortable. One of his brothers was there and he stayed with him briefly. He worked service jobs, waiting tables, retail, gas station attendant.

"I lived in various situations with roommates back when you could do it on a clerk's salary," he said. He bounced around Arlington, Alexandria, McLean, all over the Virginia side of the DC metro area.

Music continued to be a big part of his life.

"I kind of hooked up with the local culture of the rock and rollers or the people who went to parties and had beer-drinking sessions. Not really big partying, but there was that whole culture before the internet where people had to actually go places to meet people to do stuff. I was in my twenties. A lot of people in their twenties were still living with their parents—so I had the house with no parents. I had roommates. People would come over. You could be loud, they could drink beer, smoke pot. You had all these people with time and money but no place to do stuff. They were partiers but responsible, not the bad people."

He was almost thirty before he even got a driver's license when a friend had drawn a DWI and Josh was enlisted to drive him around. Besides the one brother in Virginia, he had fallen out of touch with the other members of his family, and the rift would continue for nearly two decades. His parents made no effort to contact him.

"It was just a very uncomfortable situation. The longer it went on, what do you have to say? It's probably part of my personality to be a little bit stubborn."

After a while, Josh got a job with a moving company where he met an Englishman on the crew who was in the country illegally. He started working as a house painter and got Josh to work with him.

"I don't think he did a very good job as a painter, but people thought that he was good because he had an English accent. He got away with stuff just because the way he sounded, so I started working with him. This was, like, the start of the housing bubble. I was amazed. I think part of the culture that this guy grew up in is that you put your shingle out and just start doing it. And he did it. And so I started doing it. Then I rode out the housing bubble and it was crazy for a while. People were buying houses, I would go in and paint them. They would live in them for six months and put them back on the market. And, you know, I'm back painting it again. It was crazy. I rode that into the ground and basically here we are now."

Something else was going on around the Washington metro area following the Great Financial Crisis. After all, the DC home-building and real estate scene was not affected nearly as badly as other parts of the country because government workers didn't get fired, and their camp followers in lobbying and other industries in and around government made out pretty well on the substantial table scraps of all the bailout activities.

"I watched the Northern Virginia area, the building trades, go from all white to all Hispanic without ever a passing phase where I can work there. And the reason I know that is because I lived in the house with the white people who did those jobs. They got so desperate. They were hiring white women to do those jobs. And I know because I was fucking some of them. So I want that on the record that I watched that go from all redneck white guys, white guys who were coming from Pennsylvania, Florida, staying with me and my roommates till they could get a job hanging drywall and doing that kind of construction. I couldn't get a job there. That's what was going on. You had to know somebody if you're a black. They'd hire any white person. So, yeah, I watched that with my own eyes. There's just a general work-related apartheid in the United States."

That stealth apartheid expressed itself in the service industries, too.

"You know, I'm sitting around with a group of my white friends and they told me, oh, go down to this restaurant and apply because we all just quit. We didn't get along with the manager. And you'd go down there and apply. And then later you tell them, well, I haven't got a callback, maybe the manager is racist. My friends said, well, yeah, maybe he is. And I appreciate that kind of honesty because part of the problem with racism is white people covering for white people who don't have to."

Eventually, Josh arranged the current gig as caretaker/handyman in Baltimore, an hour up I-95 from the nation's capital, and he moved into his bunkered postapocalyptic dwelling in a half-ruined building, swimming in what he called "the black undertow of ghetto dysfunction." I asked him, since he'd volunteered to live there, why he had not tried mentoring some of the kids in the hood, like the half-wild preteens we encountered on the street outside Open Works community center.

"I get the sense it's been taken over by nonprofits and you can't really do what's necessary. These nonprofits that are going to help mentor and stuff like that are largely ineffective. They won't go where they need to go. If I'm driving down the street and I see a lost child, ten years ago I would've got out and taken him to a policeman or something like that—I don't do none of that stuff. I can't because I don't want to get in trouble. It's nothing but liability. And if you can't take the hit and get back up . . . If I see a woman getting assaulted, I look the other way. I've walked past row houses where I saw a person lying on the sidewalk in front of his house. My first inclination, because of the way I was raised, was to go see if something's wrong. But then, I'm like, wait a minute, you're going to get jumped. If I had a child that was like seven or eight or nine or thirteen, he or she wouldn't be able to leave the house without saying what are you going to do before you go outside. Don't go outside to figure out what you're going to do. Before you go out, already know what you do. And if you can't do it, come back."

It was a mild evening and the Avenue was bustling with young people, predominately white people, but not by much. The common denominator, white, black, or brown, was that they all looked like people who were doing well enough, going about their business . . . constructively. The place felt safe and—perhaps strange to say—*normal* in a way that is not easy to find in the many US cities that did not benefit directly from the financialization of the economy. We were pretty talked out and woozy, though neither of us had been drinking. He gave me a ride back to the hotel in the red LeBaron, and the next morning before I caught the light rail to the airport, he delivered a copy of Neely Fuller's book to the front desk for me.

POSTSCRIPT

In a follow-up email after I'd returned home, I asked Josh what he made of the Freddie Gray incident, the spark that set off the Baltimore riots of 2015. Gray was a twenty-five-year-old black man arrested by the Baltimore police for carrying a knife. He later died from spinal injuries suffered under mysterious circumstances in the police paddy wagon on his way to the precinct house. Six police officers, three white, three black—one was a black female officer—were indicted by the state's attorney for the City of Baltimore. Two were tried and pronounced innocent, one ended in a mistrial, and eventually all remaining charges were dropped against all six officers. The Department of Justice (under President Obama and Attorney General Eric Holder, both African American, of course) declined to press federal charges against the cops.

"The trial of the six officers was a prosecutorial clusterfuck," Josh wrote back. "All the officers were overcharged and when the first two were acquitted, the cases against the rest were dropped. But it didn't matter anyway because it was all theater. Long before the trial, the city paid the Gray family several million dollars to shut up and go away. In the months after the riots, the usual agitators were screaming about racism and racial profiling, so the cops stood down, and the murder rate spiked. Let's face it: nobody wanted to be demonized, second-guessed, and called a racist for doing their job. In the end, the crime and drug problems of Baltimore may not prove remediable by throwing money at them because the actual solutions may be the practical, traditional ones that have functioned in the past such as fathers in the home, children born in wedlock, and the active respect of a common Christian culture."

I also asked him about the argument that the City of Baltimore was run by people of color, from the mayor to the police chief to the state's attorney, and many of the city employees under them, and that the sorrows of black Baltimore couldn't be attributed to white political oppression. He wrote:

"Lots of people are fooled by the 'black faces in high places' syndrome, but I'm not one of them. It's simply easier for white people to manage black people through a buffer class of 'black officials,' while at the same time it insulates white people from responsibility for the very pathologies you describe. When a problem is important and white people really want to solve it, they don't bother talking to black people first. Lincoln didn't bother talking to black people before he freed the slaves. Black pathologies are a moneymaker for lots of white people, too ... For every ineffective black government official, there are ten white people you don't see who earn way more money off urban ghetto dysfunction. 'Black officials' are doing exactly what they are required to do by the white power structure that really runs everything. You get the picture? There's a good reason why most people are unfamiliar with the works of Neely Fuller and Dr. Frances Cress Welsing."

It must be obvious by now that Josh Wickett was beyond stereotype in about every way you might imagine. He was not successful in the stock sense of someone making a lot of money, inventing some new technogadget, winning an election, or becoming a star—but he had succeeded in developing into a fully formed individual. He acted differently, viewed the world differently, spoke differently, navigated his path differently than all the clichés might induce you to believe. He was not a product of the ghetto, though he was immersed in it now. It seemed like a very lonely existence. I had to ask him why he remained there.

"With where I'm living at this stage? I thought about that and it was kind of something I needed to do," he told me. "I have a situation where I can make music. And who knows how long it lasts or if it's temporary or what. But, you know, it works for now. I have a place where I can be as loud as I want musically. Nobody is going to bother me. You talk about how important it is for you to be able to write and not be disturbed by other people. It's the same thing."

Finally, I asked Josh the hallmark question of this book, *Now what . . . ?* and to do it in writing. Since I had first met him through the written medium of blog commentary, and found his style compelling, I wanted him to answer as deliberately as possible. This is what he wrote:

I predict mass communication technology and theory will be further weaponized to the point where increasing numbers of people suffer from a Matrix-like existence; "fake news" leading the way, long on emotion, short on facts.

People will be further "freed" to become prisoners of their own passions, no matter how ridiculous. Video games, porn, recreational marijuana, "sex robots," social media of all kinds . . . "Distraction" will become the formal currency that drives the economy because in a nation where people (especially young men) can no longer count on full-time employment, the state will have every incentive to support and defend every hedonistic, narcissistic, distractive behavior in order to keep attention off themselves and their failed policies.

If and when the economy begins to clatter and wheeze like an engine starved of lubrication, the increasing chatter of pale-faced technophiles will ramp up to convince people the "next big thing" is right around the corner to save us.

When it fails to materialize, socialism will be dug up and bandied about as the solution; with the national and international socialists battling each other over its form.

To sum up, I suspect many people will simply be confused, and attempt to make their way as best they can; much like many are attempting today.

I know because I'm one of 'em.

We remain correspondents and friends.

Chapter 5

AT LAND'S END ON THE LEFT COAST

The southern end of Whidbey Island, Washington, sits above Puget Sound like a cork stopper at the open mouth of a bottle. The upper flank of the fifty-five-mile-long island faces the Salish Sea, a broad strait between Vancouver Island, Canada, and the Olympic Peninsula. Tom Thomas and Sukey Watson's ten-acre place lies at the edge of a narrow neck at mid-island, about two thousand feet off Admiralty Inlet near the decommissioned army installation Fort Casey, a nineteenth-century artillery battery guarding the sound, now a state park. It was a cloudy day in late May, and the dearth of sunlight lent an especially gloomy mood to the thick fir-and-spruce forest on each side of the highway going up the island.

I'd been corresponding with Sukey for several years, following their home-steading project. Their operation was based on an Austrian alpine planting scheme called the hugel (*hügelkultur* in German), designed for places with poor soils—and the soil on their side of Whidbey Island was hardpan, sodic soil, conditioned by centuries of salty ocean fogs rolling over the landscape. The idea of the hugel is that you create contoured planting beds (or swales) by heaping logs, branches, and other wood debris onto the surface, or into a trench scooped out with a backhoe, and fill in the space with whatever other plant debris you have at hand: compost, leaves, straw, additional soil, even old clothing, cardboard, and newspapers. Tom was getting extra organic material from a deposit of pig and horse manure eighteen inches thick laid down a century ago from the barns at Fort Casey, and also from "mining" the punky, half-rotted blowdown out of the woods on the inland side of the property.

The hugel supposedly mimics natural forest decomposition. The mounds of mixed material "cook" for a couple of years, the wood breaks down, and then you use the hugels as you would ordinary planting beds. Secondarily, the hugels are designed to catch rainwater, which is crucial out on Whidbey Island because it lies in the "rain shadow" of the Olympic Mountains to the west—meaning the cold ocean currents dump most of their rain on the ocean side of the mountains, and there's not much left by the time the weather rolls through Whidbey. The annual rainfall where they live is twenty inches a year, about equal to the sage-and-shrub high desert of western Colorado, and the scant rain only falls from November into April, so the growing season is bone-dry. It's far from an ideal situation, but it is the land that Tom happened to own in 2012 when his earlier marriage broke up and he teamed up with Sukey.

You leave the forest behind as you swing along the waterside, where the constant wind beats down on the meadows along the shore. In the distance, the mudflats lie glistening at low tide under the steel-gray clouds. Tom and Sukey's handmade house stands up a long driveway, past the open fields where Tom has been harrowing in turnips and other soil-building crops in order to grow fields of grain, mainly oats. This is for their own use, you understand: to feed themselves and their animals in the years ahead, which they reckon will be a time of hardship. Sukey started corresponding with me because she shared my view that our civilization was entering a long emergency of postindustrial decline.

I stepped out of the rent-a-car, and instantly the couple's dog leaped up to greet me. She was a beautiful, shorthaired Blue Lacy, the state dog of Texas, bred for herding hogs by the Germans who settled there. She looked like a Weimaraner, but about half the size. The dog was named Lacy—an exceptionally happy-looking, good-natured animal. Remember the dog.

Tom and Sukey call each other "Bear" and "Bird." Tom, fifty, is a sturdy, burly man of average height with fuzzy light-brown hair and a well-weathered face. He speaks in a warm, gravelly storyteller's voice with a slight western twang. He was raised in Texas. Sukey, sixty-one, is slight and delicate looking with pale skin, a Jersey girl transplanted to the Pacific Northwest with many stops in between. Both of them had endured serious physical problems. In

2016, after years of mysterious symptoms, Tom was diagnosed with a rare illness called Behcet's disease (pronounced *Buh-shays*). It was first identified by a Turkish doctor named Hulusi Behçet in 1937—an interesting detail, since Tom Thomas was an adopted child whose known ethnicity, he says, includes Irish, Turkish, and Gypsy (Roma) blood, and indeed the illness is found in greatest concentration around the eastern Mediterranean. Behcet's disease is a vascular condition affecting the skin, eyes, and lungs, with many additional complications in the central nervous system, the digestive system, and the connective tissue, inflammation being the common denominator. The disease is probably genetic, involving the body's autoimmune response. Tom walked with a cane because the disease was causing arthritic changes in his joints. Yet I spent much of the day following him around, and he did a great deal of routine heavy work: digging, lifting, and schlepping.

Sukey says their strengths and weaknesses complement each other. He has the upper-body strength, and she has the lower. Years ago, Sukey had undergone spinal surgery for scoliosis, an abnormal curvature of the spine, but she had gotten over it to the degree that she was doing fifty-mile endurance trail runs. She had also been battling severe gluten intolerance that played havoc with her gut. She had to be very careful about what she ate. The couple, newly divorced from other people in 2012, had met at a gluten-free cooking class.

We went inside, out of the wind, for lunch, a squash and quinoa pilaf with a loaf of extremely dense gluten-free bread that looked like a paving block. Sukey was experimenting with recipes for an acceptable loaf, a tough project. This version was crumbly, with mysterious flavor overtones. Inside, the house had the feel of a jewel box, all glossy wood finishes and careful joinery, with a spacious kitchen beautifully designed for food preservation and storage. At the center of the open-plan ground floor was their music recording pod, a room within a room defined by half-walls. Sukey has undergrad and graduate degrees in music, and Tom had worked as a professional musician over the years, starting in the dives of central Texas. They are avid and serious about making music and get part of their income from doing it.

Sukey grew up in a ruralish area of the New Jersey suburbs half a century ago, when there were still woods to explore near the house. She got deeply into organic gardening there; her twin sister went for horses. Her mom taught her a lot about the practical arts—at least, when she wasn't discommoded by her coping strategies.

"She had a lot of what I would call personality disorders," Sukey said. "And what I mean by that is a person trapped within herself because she always felt the need to control, but she was totally out of control. She had an addictive personality. She was addicted to different drugs, mostly cigarettes, and then finally she became addicted to alcohol. At first, I didn't really realize what was going on but as time went on . . . my mom was a freakin' alcoholic! My dad didn't know what to do about it. And she would turn into a nightmare and a monster when she was an alcoholic. But when she wasn't drunk, I did learn a lot of practical things from my mother. She taught me how to sew. She was an excellent seamstress, far better than I'll ever be. She could tailor clothes. She made me a tailored coat. And at that point I didn't realize the reason why she was making all this stuff is probably because we couldn't afford to go out and buy stuff . . . My father wasn't doing that well financially."

Her father had a checkered career. He was a chemical engineer, but he was more interested in the sales aspect than the administrative side, or the lab.

"He didn't like being in the office with, as he put it, just the corporate structure and all the bullshit that went on. He liked socializing and sitting in his car."

Tom had been adopted by a strict Seventh-day Adventist family. His adoptive father was a merchant seaman.

"He was generally never home, and when he was, he was a terror, maybe even mentally ill. You didn't want to be around him."

Tom got along much better with his adoptive grandfather and was sent off to live with him in East Fort Worth, "Bonnie and Clyde Country," he said. This grandfather was a farmer and a blacksmith. He used to read *The Iliad* and Machiavelli to Tom, who describes him as "a redneck intellectual."

Sukey and Tom's house was superbly organized, with all kinds of purpose-built storage closets, bins, shelves, and racks. One rack near the front door held Tom's collection of scythes for hand mowing grain and hay. He had become a connoisseur of scythes by using them and was partial to the Austrian type, which are "made like a samurai sword," he said, meaning the steel took the finest razor-sharp edge. He had an American-made scythe too, which he described as "garbage." These are some of the things you learn when you have "one foot in the modern age and one in what we're doing here," he said.

Sukey had some business to take care of, so Tom and I left her alone in the house for a while. Lacy the dog came along, eager for fresh air and adventure. Behind the house, a lone goose named Gomez hung out in a little corral, keeping company with two Nigerian dwarf goats, Bongo and Djembe. It was sixty-two degrees, windy as usual, rather harsh. We hiked down the driveway to Tom's workshop, a modern shed-barn he'd built himself. Much of the large interior was occupied by a Wood-Mizer portable sawmill for turning logs into dimensional lumber. The business end was a great big band saw, a scary machine that provoked thoughts of severed limbs. Tom loved working with wood, and was quite good at it, judging by the finishes and the handmade furniture in the house. But the dust from power tools was aggravating his lungs—a complication of Behcet's—and he was attempting to get around it by doing more work with hand tools that produced far less dust. One job he had underway in the shop was making teeth for a hand-made bull rake. The bull rake is a tool for gathering hay extra efficiently with every reach. He was hand cutting the teeth from the limbs of a local shrub called oceanspray (*Holodiscus discolor*), a Northwest native member of the rose family that grows up to twenty feet tall. He was using a hand tool called a rounder plane, like a giant pencil sharpener. The oceanspray wood, he said, was as hard as iron. One of the drawbacks of trying to live off the land in this region was a dearth of decent hardwood in forests dominated by softwood firs and spruces—and on top of that some new blight was killing the spruces.

Tom never finished high school, though he'd gotten an equivalency certificate later on and attended some college. But he had an entrepreneurial bent and, while still a teenager, opened a musical instruments store in the East Fort Worth ghetto, a neighborhood called Riverside, or the Crossroads.

"I dropped out, went to work, and had a business," he said. "Once I got to the ghetto, I gave up the idea to be a guitar player and I got to the rhythmic side of things. I got around black music. Then it meshed. That's what came naturally to me. Swing music and funk and any kind of boogie music. I played in all kinds of bands. I had a drum set. There was a topless bar and we used to get work—they used to pay us. We'd sneak in the back. It was called the Cellar and it's kind of infamous in Fort Worth. They'd pay us to go in there and spin records, old-school, you know, do breaks on the records. And I played along with them with the drum set. This is like in 1986. This is before anybody else was doing this. And we'd busk on the street. It was that early on. I did music on and off up until about '89, '90, and then I kind of got out of it and got a real job. I had another business. I've never really worked for anybody."

It was the early days of the computer age, and he started a business writing software: point of sales systems, audit systems, scientific software.

"It was all custom database software. One of the funkier things I wrote was paleo-dendrochronology software: tree rings."

He came to the Pacific Northwest in 2001 with a wife he'd been married to since 1989, way before meeting Sukey. Over the years, whenever he needed to supplement the income stream, he hired himself out as a "skip tracer," a private investigator specializing in locating missing persons, in particular, fugitives who can't be found at a place of residence or any usual hangouts. This investigative experience led him to doing opposition research for political candidates, something that came in handy in 2008 when Tom himself ran for a seat in the Washington State Legislature (he lost). In the early 2000s, he got into yet another line of work: as a rigger climbing towers to help set up internet communications.

"It could be anything, you know. Maybe you had to fabricate a bunch of mounts and go hang them up. There was a lot of fabrication involved. Everything was very bespoke. And I usually did it start to finish. I'd drive up to wherever it was, like, the FAA says *that* tower has got to have light bulbs. So, you've got to figure it out. You've got to get light bulbs up there, get the power up to the light bulbs. When you're running light bulbs up four hundred feet, that's a little different animal than running them up a house. And so it's an entirely different set of skills. You've got to get all the parts, and you've got to weatherproof them. Everything has to be weatherproofed, and we're talking about space shuttle weatherproofing. Because every time you go up sixty feet, the wind squares [in speed and power]. And here everything is near salt water."

Tom's interest in the subject of economic collapse developed slowly and convergently. Growing up with his grandfather meant listening to stories of the Great Depression, which hit Texas especially hard in the Dust Bowl years. He claims to be "agnostic" on the subject of how much human activity has contributed to climate change. But his interest in Earth's changing climate cycles goes way back to the sixth grade, when he wrote a report on William Herschel (1738–1822), the German-born court astronomer to King George III.

"Herschel hypothesized that when there were a minimum amount of sunspots, commodity prices always went higher," he said. "Now I got to looking at that, and I went back and I wrote this rather elaborate paper. This teacher I had was a total jackass. He didn't like my paper and he said my appendix was incomplete, and blah blah blah."

Years later, Tom was taking some time off, living off money from businesses he'd sold, and going from one doctor to another trying to get a diagnosis for his mystifying symptoms, and the now well-publicized issue of climate change reignited his interest in Herschel's observations about sunspot activity. He studied up some more. At the turn of 2008–2009, sunspot activity completed an eleven-year cyclical low after years of declining, and in March 2008, wheat prices hit a cyclical high of nearly twelve dollars a bushel.

"What was different from when we were kids is our population has grown a lot greater. As a person who's grown up as a hunter, I've seen population dynamics. I've seen [animal] populations crash. We were discussing how the eagles here are starving to death, and I've seen the population crash once before in the twenty years I've been here. We're fixing to see it again. Well, through what I would call redneck empiricism, I made some observations. You look at the problem from an economic standpoint, and then my grandfather talking about the Great Depression—which was not really a scarcity of goods, it was a scarcity of the movement of money. Well, what happens if we don't just have a scarcity of the movement of money but the scarcity of goods, too?"

Since he'd grown up in the Texas oil patch, he could also see how the industry responded to early rounds of peak oil after 2005—by finding new and more expensive ways to get the marginal producible stuff out of the ground, namely, fracking in shale plays, which Tom viewed as "kicking the can down the road a little further." His experience as a rigger working on wind turbines left him unimpressed about the potential for "alt" energy.

"For a while I was on the renewables kick and I've studied wind technology a lot. Being in the tower business, I had sites that needed windmills or solar panels. I figured out real quick what a goat fuck that was—pardon my French. I've built windmills, and I've dealt with batteries and built solar panels. And with the current set of technology in batteries, there's no way that we're going to replace any kind of the usage we have fossil fuel–wise with any of those technologies. I'm out here in the middle of an area where the wind blows all the time on the side of a mountain near the water. I'm getting eight hours of wind a day. I built a windmill that I'm creating a kilowatt an hour every time the wind blows, and I'm still having a hard enough time generating enough power every day and storing it in three thousand dollars' worth of batteries to power a few watts of radios and test equipment each day.

"Great Broadband Communications was the name of my company. We all have epiphanies in life. At this point—this was about 2007–2008, maybe '09—I beat the dead horse a little longer thinking I could maybe solve the problem. But I came to the conclusion that the problem with the current set of tools was not solvable. We used wind and solar in one form or another to power our planet for x thousands of years, but we've never powered it at this scale. The Romans, you know, they had sources of water power. And so did the Chinese. But you know what? We're burning through a fossil resource. It's like I'm going out there in that woods and I'm taking up a hundred years' worth of wood that was logged off during World War II. So I'm sucking up eighty years of the compost, essentially a fossil resource, and I'm hauling it out and putting on the garden. That's a one-shot-per-lifetime deal. It will not come again in our lifetime."

Tom was getting a picture of the complex web of dependencies that boded poorly for continuing the way of life that so many Americans take to be permanent and settled.

"One day I'm sitting here and I'm fabbing up [fabricating] something. I'm thinking, well, I'm a redneck. I can weld, and I can fix it, I can grind it. I've got a bunch of welding rods that got a little bit of wet—and once they've gotten wet they're usually screwed—and I'm out at my truck. I had this big truck in those days. We got six miles to the gallon. It was basically a platform for the welder and the crane and the snowplow. That was all it ever did, and we just drove from job to job. So, I'm out on it chucking a welding rod, trying to find one that won't stick. Shit, I've got to go to town and I've got to get my welding rods because these are all ruined. So, I go to town. They're out of stock. Oh shit, I've got to go all the way to Seattle, two hours, or have them FedExed or I'm dead for three days while I'm waiting on these parts, because stores are not going to have any more of this particular rod I need for two weeks. So, I had to call the welding supply and have it FedExed in. My big epiphany: the moment you run out of modern metallurgical tools, a lot of things that we take for granted today you can't do anymore.

"Okay, two epiphanies," he continued. "The third epiphany is back to William Herschel. About the same time, I read this weird little paper on the Gleissberg cycles by a guy named Timo Niroma, who is now deceased.[23] He was a Norwegian. At this point I was a firm believer in the whole global warming thing, sort of. And I was just going along with what I had been seeing and . . . okay, fine. But I'm looking at it all, and I read his paper and I had about four questions. I wrote him and then, holy shit, this guy answered me back. He and I started a rapport over a few weeks, a couple of months actually. He and I—and my friend Jim, at the time he was still a legislator—got this conversation going. One of my comments was: In your opinion what would happen if we had a repeat of the Gleissberg cycle that happened in the late 1700s into the 1800s? And the takeaway was, we're screwed. There is no way at this population level—even if the carbon levels could offset some of the temperature differences—there's no way that we can feed ourselves, period."

---

23 Timo Niroma, "Sunspots: From basic to supercycles," http://personal.inet.fi/tiede /tilmari/sunspot4.html (discontinued 2018).

Timo Niroma had "smoothed" the math around a great deal of data going back to the early days of modern science when the available weather information was sparser and cruder. He concluded that the longer the cycles were, the deeper the average temperature tended to fall at the end of a cycle. Niroma also used paleo dendrochronology—tree-ring studies—which Tom Thomas had, by coincidence, written software for years earlier. Meanwhile, following the collapse of the Soviet Union, a series of Siberian weather stations went offline. Tom wrote some new tree-ring software routines for Niroma to account for the missing information.

"That was actually another one of those small epiphanies," Tom said. "It was just one of those things: a little light goes off in your head. Maybe that would skew some data if you take all these weather stations off-line in the coldest part of the world. So, I started thinking about all these big things. Again, my crux is: I fear that we might start having crop failures, particularly in our commercial, monoculture-favoring agriculture. Every time that's happened in civilization, it's always ended badly. I don't care if it was the Irish in the 1840s. You know in pre-Columbian America it happened over and over again—look at the 1300s. They came into a Gleissberg cycle. They didn't particularly get cold; it just got dry. You have lower solar radiance, you get less water picking up off the ocean."

He was referring to the collapse of the Anasazi who occupied the "four corner" region of the American Southwest, for whom the great drought of 1275–1300 was the culmination of several prior incidents of drought that stressed their highly class-stratified society and provoked political and religious strife. As it happens, the droughts are recorded in tree-ring evidence. The ruins of the Anasazi population centers around Chaco Canyon in New Mexico show evidence of cannibalism in human bone remains. It may not be incidental that the founding of the Aztec capital Tenochtitlan in central Mexico happened in the 1320s, coincident with the Anasazi collapse. The word *Aztecah* in the Nahuatl language means "people from Aztlan," a mythical place of origin in the north. Exactly two hundred years later, when the Spanish under Cortés entered Tenochtitlan, Aztec society was roiling in political and religious frenzy as the population outgrew its food sources, provoking orgies of human sacrifice and cannibalism, in accordance with their religion. The horrors of the episode were recorded in the diaries of Bernal Diaz, one of Cortés's soldiers.[24]

"My conclusion," Tom continued, "is that we have a food supply that is incredibly weak. I grew up around commercial agriculture. You know what? It only takes one kick of one leg out of a stool. I don't care if it's potash, I don't care if it's oil, and I don't care if it's weather."

By now, we'd returned to the house, and Sukey was working in the close-by planting beds where they were growing artichokes and cardoons as perennials. There were already dozens of fist-size buds on the artichokes. I asked Sukey if she had arrived independently at her personal view of a problematic future as the techno-industrial era winds down.

"I had my own ideas about food security. I lucked out because I met Tom. And I met someone who shared a lot of these ideas and who had a lot of concerns about where our country, and globally, we were heading in terms of economics, in terms of energy."

SUKEY'S JOURNEY

Sukey had a remarkably rich and varied vocational life before she met Tom. Growing up in New Jersey, she got a BA in music at Rutgers, the state university, in 1978. Then she got a "full ride" scholarship for an MA in musicology at the University of Washington in Seattle, a pretty straightforward progression at that point.

"I viewed it as an *out* of the East Coast. I wanted out of the East Coast badly."

Then her life got complicated and stayed complicated for many years. At "U-Dub" (as they call the U of Washington) she met her husband, Rob. He was a pre-med student, but he veered out of that into the campus ministry. He got ordained in a branch of the Church of Christ associated with the nineteenth-century Campbellites—who, strange to relate, were against the

---

24 Bernal Díaz del Castillo. *Historia verdadera de la conquista de la Nueva España.* English: *The True History of the Conquest of New Spain.* Public Domain.

use of musical instruments in church sacraments—and he found a position in the campus ministry of Washington State University over in Pullman on the far side of the state near the Idaho border.

"I never really finished that masters in musicology," Sukey said, "because I was so utterly bored with the whole idea of studying these dark diacritical marks in these ancient manuscripts, and, like, who the hell cares, because ultimately you've got to play this stuff. So, what the hell am I going to do? And I decided, well, I'm interested in a million things. I went to the library school back at U-Dub because they didn't have that at WSU. I commuted back and forth [from Pullman to Seattle]."

At the end of her final quarter of classes for the MA in library science, she had double fusion spinal surgery.

"I was walking around in a body cast and looked like death warmed over. I think people were afraid to hire me."

But she finished her master's thesis during her recovery. And the operation was a success, leaving her free of pain to the degree that she was able to begin trail running.

"Trail running is cool. You get to run with a view, and the prize is getting to the top and getting a great view, and what better prize than to be out in the woods."

Meanwhile, her husband, Rob, grew discontented in the ministry. He decided to quit and go back to medical school, and got into the program at Saint Louis University, a private Jesuit school in that city. Sukey went with him and, to support the two of them, got a job there with Monsanto, the chemical company. She worked in the environmental health and safety division for the company that had developed Agent Orange, an herbicide used liberally during the Vietnam War—implicated in birth defects and mysterious diseases in returning US soldiers—and more recently for glyphosate (the active ingredient in the weed-killer Roundup), implicated in human cancers and lately in the widespread disappearance of honeybees. She was running the Environmental Health Information center, basically a corporate library, and writing software for the toxicologists. The company paid her to get another undergraduate degree, this one in biochemistry, at Webster University in St. Louis, which she managed to do around her regular work.

"I was a busy woman. Well, he was busy in medical school and he never liked studying at home. It was too distracting. But I really think he was probably having affairs. I don't know for sure. I mean, I'm just speculating on that. [Working at Monsanto] was such an eye-opening experience because I thought: *Oh, I get it . . . I've got this job in environmental health and safety, and it's a bunch of bullshit.* Because the main purpose of the whole division was to make sure the environmental lawyers go to DC, lobby the EPA to get X Y Z chemical off X Y Z regulatory list. It was a bit of a conflict. I was seeing what they were doing. Some people there were good souls. But no one ever discussed that, you know, this is all to cover up as much stuff as you can, yet publish the worst toxicology papers in the most obscure journal you can find. That way it's published and legal. You know, *The Laotian Journal of Toxicology*, and, I kid you not, the stuff was there. In Laotian. Yeah. Why not?"

After four years in St. Louis, Rob became a medical doctor and forthwith joined the US military to do his residency. They sent him to Fort Bragg, North Carolina, the Womack Army Medical Center. Sukey came along again, as wives were expected to do.

"Another learning experience," she said. She got a job with a fifty-mile commute each way at "a nice hospital" in a town called Pinehurst that allowed her to build a brand-new library over a three-year stint. Then the army sent Rob to Germany.

"That was really fun. Now he's an army doctor. It's payback time! He owes me three years [that she supported him through med school]. So, payback time was at this training ground where they did a thing called MILES laser training [multiple integrated laser engagement system] where they had laser wars in the mud box with a bunch of tanks and shot each other up. It's like laser tag, only you do it with big tanks instead. That was in a place called Hohenfels, Germany, in Bavaria, a great place. It was three years there. So I studied German the whole time. And then I went to the Goethe Institute in Dresden for about six weeks."

It wasn't a degree program, just study, and she was not able to get the kind of paid employment she'd worked at back in the States.

"I had some little, measly jobs at the chapel. I directed a choir and played a couple services out there in the bucolic German countryside. We lived about sixteen kilometers south of the army base. So, all my neighbors were German, pretty much. There were two Americans that I knew there from the base. I went down and talked to people in the local café. Everybody got to know me on a good speaking basis, because they all knew I was out there, and American, and they were impressed because I'm studying German and most of them didn't speak a lick of English. You'd go down to the local watering hole and they were so drunk you couldn't understand a word they were saying anyway."

Sukey was very frustrated that she couldn't do any gardening through all those years moving from pillar to post. Finally, her husband got discharged from the army and the couple returned to the USA, specifically Mount Vernon, in the Skagit Valley north of Seattle, where Rob worked as a family physician and Sukey got to grow flowers on a property that was too shady for growing vegetables. She did volunteer work for a local air pollution authority. Then she found a job as a librarian in a rinky-dink hospital that didn't have a budget to finance any library activities, but she stayed for six years out of sheer inertia. When the crash of 2008 came along, they didn't have enough money to pay Sukey anymore.

"They were hit hard. They had just gone through a huge hospital expansion like a lot of other health care institutions, because, you know, you needed the private rooms and the lawyer foyer and the healing garden. When I had that back surgery, I was in the hospital for three and a half weeks. I felt too crappy to go to a healing garden. I just wanted to go home. This is just more corporate bullshit. Of course, the CEO got a huge bonus for this. So did the other higher-ups. People like myself who weren't essential got, well, 'graduated.' So I was unemployed."

Her marriage had gone from bad to worse. Her husband wasn't even coming home nights until three o'clock in the morning, and she suspected he was carrying on with another woman. He was, in fact, and it turned out he had a child with her. Sukey and Rob never did have any kids.

"Because of this back problem, they told me not to have kids, that it would be problematic. Since I didn't want any, I didn't really care, and ultimately, I think, that's why we ended up with a divorce, because he wanted kids. His mother controlled him, and he was a wimp. And she was concerned that he would spread his seed because his seed was the ever-important seed of the family. He was the most important son, the favored son. So, I was the wedge that came between him and his mother. And she worked on him for years about all of this. Actually, he ended it. I didn't. I should have walked out on him, because we had really kind of separated. I would try to have discussions with him about what's going on in the world and he had no interest in learning anything about economics, learning anything about what's really going on in government. All he knew was medicine and nothing else. There are so many things going on in the world, and we couldn't even have a discussion."

In 2012, Rob served Sukey with divorce papers at their home while he was off at work, and that was that.

"My job had ended in 2008. I'd been like lollygagging around doing stuff and running races, being a housewife who did a lot of art, music on the side and, like, oh shit, what am I going to do now? I haven't had a job for four years."

Six months later, she met Tom Thomas at that gluten-free cooking class.

"We kind of hit it off and saw each other a few times. And then we kind of knew that it was cool. But if you'd asked me twenty years ago, I'd have to say, I need somebody who's got an education and blah blah blah. They should at least have a masters, preferably a doctorate, blah blah blah. And I got redneck instead. But it's okay 'cause life is like that, you know."

Tom had already purchased the land on Whidbey Island.

"My ex-wife and I bought it together," Tom said, "and we kind of had planned ultimately maybe moving over here. But we were not agreeing on things. I'd been having health issues since about 2008, when I ran for the state legislature, and I mean serious health issues, not just irritations. But in about 2010, it had gotten worse. And by 2011, I was having difficulty walking. One morning, a guy I worked with told me to go home. We were out in the woods logging, and he's like, 'Jesus Christ, I'm not going to work with you if you start drinking again.' 'I'm *not* drinking.' He says, 'Well, why can't you even walk in a straight line anymore. You're drunk.' 'I'm not drunk!' It was the Behcet's."

It was gray and gloomy outside, though perhaps another hour until sunset at this time of year. We'd been talking a long time. Suddenly Tom noticed that Lacy, the beautiful gray dog, had not come to the door as she always did around dinnertime. It alarmed him because she was a mature dog of unusually regular habits who always showed up around supper time if she'd happened to be outside. He left the kitchen table and went out to call her. I had some more of Sukey's experimental gluten-free bread with butter. In a little while Tom came back, saying he couldn't find Lacy anywhere. He was quite disturbed. Sukey went out to help him search. I stuck around in case she turned up at the door.

About twenty minutes later, they came back, just the two of them. They'd found Lacy stone-dead in the back meadow up against the woods. She had several large puncture wounds in her chest. You can imagine how distraught they were. For a while, not knowing what else to do, we sat at the kitchen table trying to puzzle out what might have happened. They'd found Lacy less than two hundred yards from the house. We hadn't heard any gunshots. There was a lot of fraught silence. Finally, Tom got up and went back outside.

In a few minutes Sukey and I watched him driving the mechanical back-hoe around the side of the house. Lacy was in the bucket. Night was falling now in earnest. Through the gloom, Sukey and I watched Tom dig a hole about one hundred yards down past the last of the hugel planting beds. It only took him fifteen minutes to excavate a suitable grave, place Lacy the dog in her final resting place, and cover her up. When he returned to the house there was little to do and little to say. Everybody felt so bad. We were all quite conscious of how the incident signified the conditional presence of all of us in this world of hazards. Tom eventually formed the hypothesis that Lacy might have been killed by the kicks or antlers of a black-tailed deer. She often chased them when they came around the property. Contrary to some of our sentimental Disneyesque notions, deer are large animals who will fight desperately to defend themselves when threatened and can be quite dangerous.

We all retired early, without any supper besides the bread I'd been nibbling at. Who could blame them? They'd prepared a charming little guest room for me, all wood paneled with a kind of bunk bed that made the little chamber feel like a cabin on an old-fashioned sailing ship. It took a while to actually get to sleep, and I woke up to the iPhone alarm around dawn. Tom was up, too. There wasn't much to say. We were all still preoccupied with the shock of what had happened. He went ahead of me to unlock the gate at the end of the long driveway. I got into the rent-a-car and headed off to the ferry landing twenty miles down island to catch the first boat back to the mainland.

POSTSCRIPT

I expected to stay in touch with Tom and Sukey, since we'd been corresponding long before I'd even decided to write this book and to include them in it. I also had some unfinished business with them. It was my plan to ask all the interview subjects a final question—"Now what?"—in terms of where they saw themselves and American society going in the years just ahead as the long emergency rolled out. And because of the shocking death of their dog, I'd had to end the interview abruptly without getting to that.

Sooner than Sukey had initially expected—later that summer—they acquired another dog, a rat terrier, which they named Jaeger, "hunter" in German.

A couple of months later, when I was done running around the country in airplanes and rent-a-cars, I got this letter from Sukey that pretty much answered the final question:

Dear Jim,

Tom and I have decided to put the house and additional acreage up for sale. We are moving to NE Vermont, near Cabot. [. . .]

The Navy is pushing through a huge flight expansion and it will kill our recording, music, and media business. We know we cannot fight the Navy and not only that, they are bringing in by 2020 the Marines too. They have already forced out businesses by the Port Angeles air field because they are now using this too. We have inside information that they really want the upper 2/3 of Whidbey for themselves.

We have communicated with legislatures and with personal conversations. There are no plans for buy-outs or sound mitigation. [. . .] Any funding will be years down the road and it will have to come from the county who will have to get it from the state who will have to get it from the Feds . . . That is a long chain, and by the time the lawyers get their take there won't be much to spread around even to those of us at the front of the line.

We decided to take advantage of a very hot market because it will go down in earnest once they start flying. [. . .]

We are looking for over 100 acres of timber and have already drawn up house plans which are very very similar to what we have here with a few improvements. We will have it built to the [S]heetrock and finish it out ourselves while we, gasp, live in the even smaller RV that we have now (legacy from Tom's divorce) . . . Builds fortitude and character . . .

I am glad I still have my snowshoes and heavy winter clothes . . .

It is a very depressed area but that does not affect us. We are looking forward to the seclusion. Also too, we will have more secure water and are looking for properties with on-site surface water (streams, ponds). This puts us much closer to your neck of the woods. My current mantra . . . I am not too old to do this all over again . . . *ohhhhhhmmmmmmm.*

Sukey and Tom

As it happened, they managed to sell the Whidbey Island property very quickly. Tom flew east after that and found a rural property just outside the town of St. Johnsbury, Vermont, in the region called the Northeast Kingdom. The forty wooded acres came with a house only a few years old that they intended to severely remodel. Later that fall they packed up a big truck and headed east cross-country with Jaeger the dog, Wookie the cat, the Nigerian dwarf goats Bongo and Djembe (in a travel pen), Tom's Wood-Mizer portable sawmill, and a lot of other tools. Gomez the goose was rehomed back on Whidbey Island. After a "hairy" journey that included creeping across the Rocky Mountains on icy roads at twenty-five miles per hour and subjecting themselves to the junk cuisine of roadside America, they landed at the new Vermont homestead three days before Christmas.

We still correspond regularly.

Chapter 6

STRANGE DOINGS IN THE QUIET CORNER

They call this part of Connecticut, Windham County, "the Quiet Corner," tucked away under the Massachusetts line, with Rhode Island a few miles to the east. The county roads take you through an old (for the USA) landscape of forest and farm punctuated by factory villages where the elegant Victorian redbrick mills with their once-proud cupolas stand empty and abandoned, wistfully evoking a bygone America of people who had some reason to get up in the morning. A striking sample of the folks I saw on the streets in these places looked disheveled, pushing shopping carts along the sidewalk, far from wherever the shopping carts originated. You had to wonder: What has happened here?

The Quiet Corner gives off an eerie vibe, like one of those New England backwaters in the 1930s horror stories of H. P. Lovecraft—where the sun goes down twenty minutes early and the furtive denizens who scuttle through the grim streets all have eleven fingers, and the wind carries necrotic odors. I went down there to meet up with Rob Freeman, who has corresponded with me for about fifteen years. I'd hear from him via email a few times a year. He was always polite, and he always had something to say that was outside the box and unusual. Rob is a white nationalist. He had been reading my long-running blog, *Clusterfuck Nation*, and he thought I would be interested in what he was doing.

Well, I *was* interested because white nationalism is another facet of the identity politics that the nation has been choking on for years. White nationalism is arguably the most unappetizing of the many tensions in these identity politics because it reminds people of the American original sin of slavery and its stepchild, Jim Crow, and therefore induces mortifying shame among the educated and well-intentioned especially. In short, it is beyond the pale. Yet, it is with us whether we like it or not, and it requires our attention.

For the record, I'm not a white nationalist. I'm a born-and-bred American of Jewish ancestry, non-practicing—I'm more interested in the Boston Red Sox than the Talmud. I consider myself a Caucasian, though some in Rob Freeman's circles might dispute that. (We'll get into that further.) I've been a registered Democratic Party voter for nearly fifty years, though I did not vote for Hillary Clinton in 2016 (nor her opponent—I wrote in David Stockman). I am, shall we say, a disaffected, old-time, Jack Kennedy sort of Democrat. Those days are long gone, of course, and we are a very different country now than we were in the 1960s, and the Democratic Party is not what it used to be when I first signed up for it.

In these years of amped-up identity politics, I have written many blog posts about the racial, ethnic, and sexual upheavals that have so grievously divided the nation. These included the Trayvon Martin killing; the Ferguson, Missouri, riots; Black Lives Matter; the Dallas police massacre; the Charlottesville dustup; the Brett Kavanaugh hearings; the social justice uproars on campuses everywhere; and more. I am required to process news and opinion from all quarters, including places outside the "Overton window" of acceptable thinking.[25] The pressure to conform to received thought is considerable these days, what with Twitter mobbing and social media de-platforming and de-monetizing, which means that apostates to the *bien pensants* can be treated very harshly, even deprived of a livelihood if their work is web based. I refused to be squeezed into the template of pre-cooked, politically correct thinking. Some readers might be "offended" that I'd even undertake to investigate Rob Freeman's world. But ignoring it will not make it go away. So that is what we are going to be looking at closely in this chapter.

---

25 The Overton window is the range of ideas tolerated in public discourse. It assumes, as well, a range of ideas outside the window that are not tolerated. The term is named after political scientist Joseph P. Overton (1960–2003).

AN AMERICAN LIFE . . . WITH RUSSIAN DRESSING

After a pretty rough childhood and decades of personal peregrinations, Rob Freeman landed in the Quiet Corner town of Putnam, on the Quinebaug River, running a taxicab company. He came to meet me at the Comfort Inn in one of his cars on a raw, early spring afternoon and took me on an orientation tour of the town. At age fifty, he's muscular, with a weathered, ruddy face and fluffy brown hair, wearing well-worn casual clothes. He's married to a Ukrainian gal named Anna, and they have a grown-up daughter, age twenty, who is working as a bartender and has a place of her own. Rob paid for her to go to bartending school under the theory that hands-on work is salutary for young adults. He expects she'll move on to more substantial work. Before that, she was homeschooled.

"I came to the conclusion that to be a white nationalist, the best thing you can do is study math and physics and maybe foreign languages and to be a homeschooling parent and raise children from a young age who are really good at math," he said, as we motored through a neighborhood of substantial old houses, many in need of renovation. Rob himself likes to do math as a hobby "for cognitive enhancement." These days, his favorite math workout is how the transcendental number "e" comes into existence. "It's like pi," he attempts to explain. Of course, he was talking to somebody who flunked both basic geometry and algebra 1 in high school.

For years he has written to me about the tribulations of running his cab company. The majority of his passengers are poor folk on Medicaid going to doctor's appointments, or to kidney dialysis, or to methadone clinics. Their taxi fares are paid by the state of Connecticut via a third-party company that handles all the state's billing. And lately, that third-party company had been mysteriously delaying its payments to vendors such as Rob for weeks at a time, making it difficult for him to pay salaries and revolving expenses. Most taxi companies pay their drivers a percentage of the take—generally 60 percent—and also make them pay for the gasoline they burn. Rob thinks that's foolish. He puts them on salary and pays for the gas because it's a tax-deductible business expense for him. The result, he says, is fewer traffic accidents because his drivers do not take excessive risks to boost their income.

"The dispatchers are not allowed to tell them to hurry up. They're not allowed to stress out at them. Everybody has to be nice to everybody or you're fired. If anybody comes into this organization who is not nice, I bounce them. I keep everybody as low stress as possible. And so they are safe, because I'm a Buddhist."

He has ten employees. As it happens, he hires people of diverse backgrounds. His longtime dispatcher is a Puerto Rican woman. He's tried to encourage her to homeschool her kids, and gave her copies of the old *McGuffey's Readers* to take home, but he said she wasn't interested. He's hired immigrants from sub-Saharan Africa to drive for him. He fantasizes about becoming the public transportation chief for Windham County, which is very poorly connected to job centers in New London, Providence, or Boston. Boston is only an hour away by train, but train service was discontinued long ago.

"A Windham County person would kill for a sixteen-dollar-an-hour job," he says. "All kinds of jobs in restaurants and motels and the colleges, the Coast Guard Academy—there's tons of jobs and tons of money in New London County: seafood processing, they build submarines. So it's like North and South Korea. And so that's why Windham County has the worst heroin addiction in the state. It's economically isolated. I could open a computer school and teach people computers, and they could work in Boston if they had transportation to Boston. But they'll bring people in from India to do the computer stuff."

Putnam was a mill town surrounded by farms. But farming in these stony river valleys just can't compete with other regions of the country. The factories, which once made uniforms for Union soldiers in the Civil War, are all quiet now. Putnam also happened to be the unlikely setting for the establishment in 1933 of the All-Russian Fascist Organization led by the anti-communist Anastasy Vonsyatsky, who was arrested by the FBI in 1942 and charged with cultivating secret contacts with agents of Nazi Germany, after which his party evaporated—a strange footnote in history.

Before the offshoring of American industry started in earnest, the unlucky town was badly damaged in August 1955 when two hurricanes, Connie and Diane, struck within a week and flooded the Quinebaug River.

Among other losses: the railroad tracks were washed away. When the 1970s rolled in, it was all over. We drove around the town, or what is left of it. Like a lot of de-industrialized places in New England, Putnam became a stop on the antique hunters' circuit—the emptying of barns and attics being a short-term work-around for economic desperation. Now, that's grinding to a halt, too, as most everything of value has been discovered and sold off.

"Living in a bleak place makes you either a heroin addict or a hero," Rob remarked. There are other outcomes as well.

He told me the story of an acquaintance, a local real estate developer named Greg Renshaw, who was shot in the head one August night in 2016 at age fifty-five. Renshaw owned the Cargill Falls Mills, an ensemble of old redbrick and dressed-limestone buildings. He was attempting to create a mixed-use project of apartments, offices, and retail in the complex. The old factory buildings were located beside the river, naturally, and part of the redevelopment concept was to install hydroelectric generators there to run the buildings. But the town government had caused the project to shut down for nearly a year as they attempted to assess the effect the proposed hydro installation would have on the scenic falls. Rob theorized that Renshaw had borrowed money from the New England mob and the delay put him badly in the hole, missing payments to people who were deadly serious about collecting. Years later, there are no suspects in the case. Rob took me inside Renshaw's first-floor office suite in one of the mill buildings. The door was unlocked. After two years, the place was still stuffed with fine old furniture, a lot of electronic office equipment, and quite a few large nineteenth-century oil paintings, plus antique lamps and other stuff. I couldn't fathom why it hadn't been carted off by junkies and thieves. Rob did not have a theory on that. The mill renovation project is dead too, of course. I couldn't wait to get out of there.

It was getting on supper time. We settled into a booth at the Courthouse Bar & Grille ("Arresting food and spirits!") in the "antique district," which was a rebranding attempt for the old center of Main Street. The bar was hopping on a weekday night with a mixed crowd of tattooed young blue-collar types and local professionals in office togs. We ordered beers and food and he told me his life story.

He says he had "a terrible childhood" in Glens Falls, New York, an upstate mill town affected by the same historical forces as Putnam. "It was like *Lord of the Flies*," he said. "The other kids were bloodthirsty, and I was a little weird and eccentric and later in life I realized that because I'm into books and studying math and learning Russian and playing violin, all that kind of stuff, I was, like, a mutant. So, they came after me for that."

His mother threw his father out of the house when he was eleven. She hired babysitters who gave Rob beer and booze. By thirteen, he was binge drinking with friends, including one of his former tormenters, who had taken him under his wing and introduced him to weight lifting.

"I'm so lucky to be alive. I'm so lucky to have survived my childhood. We would go out in the woods and drink a lot and go down by the Hudson River. One time it was the Satan rock stuff, Iron Maiden, the devil worship thing. So, they get the idea: let's sacrifice Freeman into the fire—because we had a fire. And I was like, you know what, they would really fucking do it. And I slipped out of those woods. These were my peers."

At fifteen, his mother had enough of him and threw him out of the house when the police brought him home for sleeping off a bender on a stranger's porch. His father was a flight instructor, teaching people how to fly small planes, and he had relocated to New London, Connecticut, so Rob went there. He went to the public high school—"a nice one"—and very consciously undertook to clean up his act. He scored well on his SATs and took Advanced Placement courses. He developed an interest in learning Russian and he went to Norwich University in Vermont, a private military college that offered Russian language studies.

"I was there long enough to get hazed and I didn't punk out. By the end of the semester, I wanted something more academically challenging."

At Norwich, he had been required to enroll in the Army ROTC. He left Norwich and enlisted in the regular Army Reserve. He took basic training at Fort Leonard Wood in Missouri—"Fort Lost-in-the-Woods Misery," the soldiers called it.

"The guys got along great. White guys, black guys. But for the girls, black and white, it was a race war all the time. The women's basic training was separate from us, thank God. Now supposedly they're mixed. That would be horrible. Basic training with guys was great. For guys, basic training was fun. The girls were, like, oh my god, it was horrible. The drill sergeants were fucking the recruits, and these black drill sergeants preferred the white girls. The black girls were jealous. And that's why they would fight. The girl recruits would fuck for favors. And the sergeants were in a position of power. This is one of the dynamics about a mixed-gender military that seems perfectly something you'd expect. And yet the authorities pretended that that wouldn't happen."

He described the men's side: "[T]here was a kid who was obviously mentally ill, had no business in the army, and he would do anything to cause a scene. One of the things he did was he called a black recruit a 'porch monkey,' like trying to spark a race war, and all the white kids told the black kids, 'He doesn't speak for us, we don't want any racial cause [i.e., black/white strife] and we'll take care of it.' And they ended up giving him a soap party, beating him up in the middle of the night. You put soap in a sock or a pillowcase and you beat the crap out of this kid, so we took care of it. We didn't want a race war, not even the slightest hint of it. We got along great because we were all business, we just wanted to get this shit done and get out of basic training and move on. So then I went to the Defense Language Institute at Monterey, California. It was so beautiful there. Well, I could never afford to live there. And I got to live there for nine months."

He was getting very proficient in Russian. It was 1992. Rob supported conservative Pat Buchanan in the Republican presidential election primaries. His inchoate adolescent angst, once centered on heavy-metal music, was morphing into a political view aligned with the Right.

"The teachers liked to use me because they knew I would get the students angry with my politics. And I was much better at Russian. They would want to argue in English in order to yell at me, and the teachers would say, 'No, you've got to say it in Russian.'"

Rob did two additional months in Voice Signal Intelligence Training at Goodfellow Air Force Base in San Angelo, Texas. After all that, he was able to enroll in Connecticut College back home in New London, with a break on tuition as an army reservist, continuing his Russian studies. Midway through his undergrad program at Connecticut, his father and sister were killed in a plane crash near Syracuse, New York. His father had been taking his sister back to college there.

"I was working and I came home. I didn't have an apartment at the time. I lived with my father. And there were a zillion messages on the answering machine. But the phone rang as soon as I walked in. It was my friend's parents. They said, 'Come see us.' They didn't tell me what had happened. I went to see them and then they told me."

Of course, it affected him emotionally, but the tragedy impelled him to accelerate his college studies, and he graduated six months early. He didn't know what to do, but he'd been wanting to visit Russia, and now he made that happen. He was still in the Army Reserve. He got permission from them to go. He flew to Moscow and found lodgings there.

"I had a top-secret security clearance for military intelligence and I was living with a Soviet naval intelligence officer. He was retired, but he spoke English. I wasn't sent as a spy, by the way. I was just told to go hang out."

He worked a few hours each day teaching English. He was twenty-five. Moscow, in the mid 1990s, was an exciting place for a young man. The Russian people had only been out of the Soviet yoke for a few years. Liberation and danger were both in the air. The Americans, led by Harvard economist Jeffrey Sachs, were attempting to "help" the Russians reorient their economy along capitalist market lines—and incidentally assisting the fledgling oligarchs in looting what remained of the country's industrial base to amass their fortunes.

"It was tough on them, but they survived okay," Rob said of the ordinary Russians who didn't get to be oligarchs. "They were fine. They still had themselves. I could LARP [Live Action Role-Play] right into it.[26] They knew I wasn't Russian. But because I spoke so well, and I was white, they accepted me. I didn't find any girlfriends. I mean, I fooled around with a few girls but I didn't find a steady girlfriend. The thing was, I couldn't support them. They wanted you to marry them, and you would have to pay for them. And I was like, hey, I can't get a job that pays well enough. I learned the word

---

26 Live Action Role-Playing (LARPing) is a game where the participants physically portray their characters; it often refers to masquerading within real society.

obyespechivat: to provide for someone. And so, when I was with a girl and it got serious, I would need to say I can't support you financially. I can't get a job. They were very pretty, and very virtuous, very nice. They would have made great wives. I just didn't have the mojo to support a real woman."

He wanted to stay longer, but his visa expired. I asked what especially appealed to him about Russian culture.

"That people get really good at one thing. Like really good at violin, really good at math, really good at gymnastics. We're all just dilettantes here in America. That's the most important thing I learned in Russia. I took violin lessons from a Moscow Conservatory teacher. I was starting to get good. I would have become a competent violin player if I could have stayed in Russia. That was such a regret that I had to leave. And then after that, they liberalized the visas."

He returned to the USA in late 1995 and settled briefly in a Boston boardinghouse. Two of his fellow boarders were Chinese immigrants, one of them an artificial intelligence professor from MIT, another a physicist earning a PhD at Harvard. They impressed him with their supernatural diligence. Then there was a third guy, an American trust-fund kid who sat around smoking weed and playing video games all day. He ran up $600 in long-distance charges on Rob's landline phone—this was years before the now ubiquitous cell phone. Rob was living on mac and cheese and pinching pennies, and the bill pushed him over the edge. He left the place rather than confront his housemate. "I chickened out," he said.

It was the winter of 1996. He fled Boston for a temporary gig back with the military in New York City. He was assigned as a Russian translator for a US customs agent under the Department of Defense's Joint Task Force 6 (now renamed Joint Task Force North). Its mission was to fight the US "War on Drugs," in particular Russian mafia activity in the New York metro area. His boss, one Joel Campanella, had been a NYPD officer for twenty years before moving into the customs office. New York City was classified by JTD-6 as an HIDTA: High Intensity Drug Trafficking Area.

Rob worked out of an office in the World Trade Center. He doesn't remember which tower, North or South, but it was the second floor. He bunked in Fort Hamilton, Brooklyn, under the Verrazano Bridge. His main assignment was to assist in translation at the trial of the gangster Vyacheslav Kirillovich Ivankov, also known as Yaponchik ("Little Japanese"). During recesses in the trial, Rob "hung out" with the Russians, in particular Ivankov's girlfriend and son, honing his Russian language skills. That brought Rob to the attention of the FBI and, soon after, the Army JAG [Judge Advocate General's Corps].

"And you know who interrogated me? James Kallstrom, the special agent in charge of New York City FBI at the time. He was doing the case. And he's like: 'Who's this kid talking to all these Russian gangsters in Russian?' He pulled me into a wood-paneled telephone booth in the federal courthouse in Brooklyn, and raked me over the coals for about five minutes—and I was like: 'Just call Joel!'"

The army pulled Rob off the trial while the JAGs investigated.

I asked Rob if he was already involved in what would come to be called alt-right activities.

"No. But I was starting to see things. Like I noticed all these Russian-speaking organized crime guys were actually Jewish. Very few of them are ethnically Russian—"

"Jews can be ethnically Russian," I said.

"Yeah, well, in Russia if you're Jewish, your passport says 'Nationality: Jewish' not 'Nationality: Russian.' They'd make that distinction there. Or, if you're a Tatar, they'll say you're a Tatar."

Meanwhile, the JAGs decided he hadn't done anything requiring prosecution or discipline and dropped the matter.

"They let it go, thank God. That would've been interesting to get court-martialed. I was invited to do another tour [of duty], but while I was in New York City I took a computer course in Novell Networks and began working in the computer field."

He left the city in October and lived briefly in Old Lyme, Connecticut, with two college friends. Awaiting his final Army paycheck for the New York City gig, he went to visit his mother in Glens Falls, where, by a strange happenstance, he was introduced to a newly arrived young Ukrainian immigrant named Anna Zubkova, who had been born when Ukraine was still part of the Soviet Union. Things got serious fast.

He induced Anna to move to New London, his old college town, where he felt comfortable. They got married there. Anna got a college degree and was offered a full scholarship from the Stetson University College of Law in Tampa, Florida. Once they got down there and settled in, the school rescinded her scholarship. It had been granted based on her returning to Ukraine.

"You can't have the scholarship because you have an intent to stay in America," Rob attempted to explain their reasoning. "Well, what's the point of getting an American law degree if you don't have an intent to stay in the country? But they're like: 'Since you got married, you can't have the scholarship; you destroyed your status by getting married.' They really were mean. I don't know what they thought she was going to do in Ukraine with an American law degree. The American Bar Association is not recognized in Kiev. They were tough on Anna. And I had to fight to keep her in the country."

They were already down in Florida, where their daughter was born. A year later, Anna got into Suffolk Law School in Boston with no strings attached, and they moved back up north. Rob got an IT job. They found a place in the ethnically mixed Dorchester neighborhood. Rob describes living there as "My red pill moment."

"One time I had a week paid vacation," he said. "And during that week I hung out in the apartment. While I walked around outside, I noticed that everybody else had permanent vacation; nobody went to work. They got free housing, free food, free everything, and I was barely making it. We were paying fourteen hundred dollars a month in rent."

It was the age when internet chat rooms were coming online in earnest and proliferating. Rob started hanging out in one frequented by members of the so-called National Alliance, "Basically what they would call a neo-Nazi organization," he said.

"You were that pissed off?"

"Yes. I started a real-life unit. I was never the unit coordinator, but I did activism and you can look it up."

The National Alliance was an organization founded by William Luther Pierce (1933–2002), a physicist who penned the notoriously racist novel *The Turner Diaries* (under the pseudonym Andrew Macdonald), in which Jews, gays, and nonwhite people are exterminated following a second American revolution. At the time of his death, the National Alliance was bringing in more than $1 million a year.

"My comrades and I would dress in business suits and show up at the Framingham Human Relations Commission, and say, 'What about white people?'" Rob recalled.

This, and other activities, sometimes led to violent clashes with their opponents.

In 2002, the *Boston Globe* did a story on Rob and his racialist agitation. He refers to it as a "doxxing" story.[27] As a result, his daughter was kicked out of her Waldorf School in Lexington.

The 1998 movie *American History X*, about a violent white nationalist played by Edward Norton, had also made a strong impression on him.

"*American History X* was like: how to totally destroy your life. That movie got a lot of young men to destroy their lives. And so my intellectual contribution in those days was: Don't copy *American History X*! We got to have our own model, we've got to find our own way. Because the question is how do you *be* a white nationalist. Like for example, who are the real tough guys? '*I'm gonna kill people . . . I'm going to beat people up.*' They ended up repenting and going to the other side. I mean, that's terrible. It's better to be a soft white nationalist who never betrays his cause than a tough guy who turns traitor. So, I said, 'Look, we should have a softer personality. You've got to be a decent human being. You can't go around saying: *I'm going to kill people.* You can't do that. You know that's just bullshit. You've got to be a nice person because white people are nice, okay. And we can't live with ourselves if we're mean. The trap that our enemies try to get us into is to be mean so then we'll feel sorry and we'll repent, right? Don't fall in that trap. Be nice.'"

"Did you at the same time harbor any thoughts about genocide against black and brown folks?" I asked him.

"No, no!"

"So, what are you then? A separatist?"

---

27 From *dox*, abbreviation of documents, publishing personal information about someone—address, phone number, employer—in order to make their life difficult.

"I wanted to live in a world where you didn't see a white girl walking down the street with a black boyfriend," he said. "That really made me crazy. I admit that, and you can make fun of me for that. I admit that I don't like that it breaks my heart, it tears me apart inside, and I saw it day after day after day after day. This is where I learned to step outside myself—*Look at me being angry*, you know—and became a Buddhist."

DEEPER INTO THE STRANGE

We reconvened the next day at Rob's house, an unassuming vernacular sixties home on the semi-rural edge of another town ten minutes from Putnam. He took me down to his basement man cave, a combination office/workout room where he practices yoga and qigong, a slow movement discipline of breathing, stretching, and meditation. There was also a martial arts heavy bag hanging from a joist.

I sensed that he trusted me to be fair about telling his story and presenting his ideas. By that I mean not treating them timidly or disingenuously out of fear of offending various groups and parties. Nor was I much interested in debating him; rather, just hearing him lay out his beliefs. Rob was under no illusions that he represented anything but an unpopular point of view. I am functioning as a reporter here and bringing you a report of what I found—and doing it in a historical moment of high political tension, when all sides distrust and even despise their adversaries.

I resumed the conversation by asking where he thinks he got the idea, or the feeling, that white people should not mix with nonwhite people.

"It's instinctual," he said. "It's your gut. It wrenches your gut to see white women with nonwhites."

"Do you believe that it's something coded in human behavior?"

"Racial identity? Of course. Yes. I think it's absolutely my DNA talking."

Where we'd left off the night before, Rob had turned thirty living in Boston. He was now a full-grown man, with a wife and a child. The part of his brain that processed judgment was fully developed. I asked him what conclusions he had drawn about the society he was born into.

"I applied what I learned in Russia," he replied. "I noted that Americans were anti-intellectual dilettantes. You know, it's hard to find a good violin lesson around here. It's hard to find someone who's interesting to talk to. It's hard to find anybody who reads books. The people who do—those Chinese men I told you about before—they made fun of Americans for being so dumb. Those guys know math. Everybody in the physics department is Chinese or Russian or Indian. And I took that to heart. I came to the conclusion that to be a white nationalist, the best thing you can do is study math and physics and maybe foreign languages, and be a homeschooling parent, and raise children from a young age who are really good at math. And the way to get them in that is: you want to train a child in the psychological 'flow' state where they are totally calm, get them to actually concentrate on something—building a model airplane, or coloring a coloring book, or playing a piano. But you want to get that flow state, and you want to get them to be able to teach themselves and sit quietly at a table and work out math problems or read a book for hours. Anyway, that's the best thing we can do for the cause."

"How do you feel about being in a culture that says that you are both unintelligent and monstrous for believing the things you believe?" I asked.

"It gives me energy. Being persecuted gives you power. I have studied *The House of Rothschild* and I read the first volume by Niall Ferguson and I love the Rothschild story.[28] They were in the Jewish quarter in Frankfurt. Germans did not like the Jews going way back. They had *Judensau* [images of Jews in obscene contact with a large sow, a female pig] and it shows Jews—one of them is sucking at the teat of the sow, one of them is eating her poop, and one of them is drinking her urine. And this is how Rothschild, you know, when you look out the window, like, *this is how the Germans feel about us.* And Jews were prohibited from commercial activities, were restricted because

---

28 Niall Ferguson. *The House of Rothschild* (Book 1). New York: Penguin Books, 1999. Paul Johnson writes: "[T]he Rothschilds are elusive. There is no book about them that is both revealing and accurate. Libraries of nonsense have been written about them . . . A woman who planned to write a book entitled *Lies About the Rothschilds* abandoned it, saying: "It was relatively easy to spot the lies, but it proved impossible to find out the truth." Paul Johnson. *A History of the Jews*. New York: Harper and Row, 1987.

if they were allowed to run free they would overwhelm the Germans and everybody knew that. Everybody knew that Jews were so good at commercial activities that they would overwhelm the Christian business. It was sort of like affirmative action for Christian businessmen."

"That's strange," I said, "because if the Jews were admired for being good at things, for being skilled and competent, why were they reviled?"

"Because they were dealing in alcohol. They would traffic in young women. Whatever way you could make money that was a little bit of a dirty way to make money, they would do that. People understood that Jews got rich on things that caused negative externalities."

"Alcohol wasn't hard to get from other Germans," I said. "And Germans were hardly against drinking."

"If you look at the beef the Cossacks had with the Jews—the Cossacks didn't like Jews. One of the reasons was alcohol trafficking and the other one was snatching girls and selling them to the Ottomans."

"How do you know that's for real and not just some slander?"

"We'd have to go back in time," Rob said. "We'd need a time machine."

"Then there's the old *blood libel* idea that the Jews take the blood of children and make it into matzos," I said. "Do you buy that one?"

"I don't know. I don't know what to believe. But I have this book by Ariel Toaff—"[29]

"Because we're having this conversation, and you know I'm Jewish by birth," I said, "do you impute any bad motives to me?"

"No, no."

"Why wouldn't you?"

"I know you," Rob said. "The best I can say is that I don't perceive that from you. And I've known you for, like, fifteen years. Personally, I like Jewish people."

"You must have known plenty of them living in the northeastern USA."

"Well, I learned Russian from them. But I don't like what powerful Jews are doing to us—people that say, 'I'm Jewish and I'm doing this anti-white stuff.' They admit it themselves. What am I supposed to believe? The people

---

29 A. Toaff. Il Mulino. *Pasque Di Sangue* (Passovers of Blood). Bologna, 2007.

who are fighting for open borders were Jewish organizations and Jewish money, and man, they got their way. You know they bowled us right over and got what they wanted."

"How do you account for that?" I asked. "The 1965 Immigration Reform Act was fifty years ago. How did *goyische* WASP America get steamrolled on that?"

"The prosperity," Rob said. "There was such a bubbling up of prosperity, money, you know, that it made the culture kind of crazy. I think that the ideology is enforced by what I call *paycheck liberals*. In your day, people were shoemakers, maybe owned a hat company, maybe owned a store. They had to have trade secrets, they had to be clever. Right? Now the last middle-class job in America is a paycheck liberal like, you know, a social worker."

"Some bureaucratic functionary?"

"Right. And so, like, for example, in Boston the alt-right organized some kind of silly civic nationalist demonstration. And five thousand paycheck liberals showed up to protest them. If you look at these people, they all have jobs either as schoolteachers or bureaucrats of some kind. All the Left that comes out to oppose us—they all get paid to be liberals, and they're not very bright because their job doesn't require them to do anything except agree with liberalism. But the one thing they know how to do is bully the conservatives, and the paycheck liberals bullied the alt-right into existence. We noticed that a Jewish person will tweet on one thing, you know, 'White people are bad, white people are racist.' And then in another tweet they'll say, 'Well, I'm Jewish and that doesn't apply to me. I'm actually not white.' So they're tactically white when they bash it, but they never say anything positive about white people. And they say positive things only about their Jewish identity."

"But the alt-right thinks that Jews are *not* white people."

"You are. But just for some reason you hate us. When people say, 'I'm not white, I'm Jewish,' am I supposed to *not* believe them?"

"Isn't it tactically a bad idea to even refer to Jewish people as not being white when they obviously are?" I asked.

"Jews divorced us. We didn't divorce them. Jews were accepted as whites and then they said, 'No we're not white, we're Jewish. We're better, we're more moral and you guys suck for your racism.' Jews initiated the split, not us."

"Do you think that there's some kind of malevolence behind the behavior of the Jews in America?"

"There's religious Jews who generally don't seem to get involved in the politics, and they say that fighting anti-Semitism and racism is, like, a fake way of practicing Judaism. And if you want to practice Judaism just practice Judaism, right? And that means those guys, maybe they get involved in some financial shenanigans. But, I mean, if Jews just did financial shenanigans but didn't do this political shit we would be, like, okay that's fine. You know, they make some money and do some moneylending, whatever. It's because they open our borders and encourage our women to turn against us. That's what really hurts us. That was going too far."

"How do women and the women's movement fit into this?" I asked.

"Well, you know, American men, they kind of got beat. We've actually figured out feminism. We're going to destroy feminism with an understanding of female psychology, just by being alpha males. The only reason feminism succeeded was because men did not understand how to deal with women. We tried to please women. We believed that women were intellectually and morally better than us, right through the '80s and '90s. That was a mistake. Women don't want us to be like that. They want us to be assertive and dominant and alpha silverbacks. They want us to be the alpha wolf. A woman will crave an alpha wolf all her life and pass over all the beta boys. Feminism is terrible, but I think it took advantage; it was like an opportunistic infection that took advantage of male weakness."

"Why did American men surrender?"

"We didn't know any better. I was living a bunch of lies all my life because I just didn't realize they were lies. I had to clean them out. Like the idea that women were smarter and morally superior, and we should look up to them instead of them looking up to us. Or that equality could be possible. That was all stupid and crazy. And thanks to the internet and the alt-right I learned otherwise. I learned actually how to be a proper human male."

"What do you and your homeys in the alt-right make of the post-Ferguson Black Lives Matter movement?"

"Oh, it helps us a lot," Rob said. "We're not particularly angry at it. They attack leftists, like the Starbucks thing. Starbucks is a leftist corporation, and they're getting attacked for that incident in Philadelphia.[30] And that guy at Hotep Nation, a sort of black nationalist who's friendly to the alt-right.[31] He went into Starbucks and said, 'Starbucks is racist, give me a free coffee!' And they gave him a free coffee. And he filmed it, and then he put it on YouTube. And he got on Fox News. He was just trolling. So, we're in a period where whites can just step aside because the blacks are factionalizing, right? The Hotep guys are saying that whining about racism forever isn't getting us anywhere. It doesn't do anything for us. It's a mess of pottage.[32] Right? And the Jews are factionalized and they're fighting within themselves, so we don't have to. We're not about hassling other races in any way. We're about just working on ourselves. But if you're pro-black, how come you don't say anything about high-fructose corn syrup? Giving a kid high-fructose corn syrup and putting them in front of the TV is the worst thing that happens to any child in America. You should be trying to get black parents to have a mom and a dad who feed their kids healthy food and don't sit in front of a TV, have them do other stuff. If you're really pro-black, that's what you would be advocating instead of lecturing about how evil white people are."

"What about the dustup in Charlottesville, Virginia?" I asked, referring to the August 2017 *Unite the Right* march that resulted in the death of a counterprotester, Heather Heyer, when a car driven by one James Alex Fields Jr. ran into a crowd, leaving nineteen other people injured.

"This guy, a professor at North Carolina State University named Dwayne Dixon, bragged on Facebook that he pointed his gun at that Dodge Challenger and scared the kid. And then the kid went and hit the crowd. Dwayne Dixon's a white guy. He's an Antifa and he's a paycheck liberal. Paycheck

---

30 In May 2018, two black men were arrested for trespassing in a racially charged incident at one of Starbucks' Philadelphia stores. The coffee chain announced it would close 8,000 of its US stores for a day so staff could undergo racial bias training.

31 Bryan Sharpe's website, Hotep Nation: https://www.hotepnation.com.

32 Biblical reference: A mess of pottage is something immediately attractive but of little value, taken foolishly and carelessly in exchange for something more distant but immensely more valuable.

liberals have no vision of a better future; they have no positive vision of what to do. I have a positive vision. Make me the leader! I'll do mass transit and homeschooling and yoga."

"How much of the strife that we're seeing now between the various racial and ethnic and religious factions in America is due to the economic disintegration of our country?" I asked.

"Oh, yeah, that definitely helps. I saw that coming. That's the biggest red pill because, I mean, money was able to grease everything. But it was unsustainable. So now the gears are grinding to a halt."

"Do you think that blacks are at a genetic disadvantage?" I asked.

"Well, they've evolved in a different climate. That said, I think most people are not getting anywhere near their intellectual capacity because public schooling is terrible and TV for children is horrible. Our culture is shit. So, I think there's a lot of room for improvement of everybody, including black people. If they did homeschooling, you could actually get black children to be much better educated and behave better."

"What if there's only one parent and it's a fifteen-year-old girl?"

"If the parents can't do it, I guess you're out of luck. Because we can't do it for them, and we don't want to do it for them."

TINFOIL HAT LAND

People who subscribe to marginal ideas, especially certain conspiracy theories, are said to be wearing tinfoil hats—like an antenna for picking up broadcasts from an alternative universe. I felt an obligation to delve into this realm of odd beliefs with Rob.

I started by asking him about the theory, popular in alt-right literature, that Adolf Hitler was a British secret agent. Yes, I know, just saying that surely provokes derisive laughter and raised eyebrows. It is nonetheless a popular meme in the alt-right folio.

"He threw the war just like Max Schmeling threw the last fight against Joe Louis," Rob replied, referring to two heavyweight boxers of the 1930s, "and he [Schmeling] got a Coke distributorship."

"It's really hard for me to believe that Hitler was a secret British agent," I said. "You're a smart guy. What makes you think that?"

"Well, I've read the history of intelligence. MI6, CIA, KGB. And through this research and study, I learned. Who killed JFK? It was Allen Dulles."[33]

I didn't follow him down that rabbit hole.

"What about Hitler?" I said.

"I think that he got turned into a deconstructed person at the Tavistock Clinic in 1911," Rob said, "and he was sent as a kind of weapon against Germany, to take over and be the inside man, a mole, right from the beginning."[34]

The popular alt-right book *Hitler Was a British Agent*, by Greg Hallett, connects this Hitler story with occult theories about the Rothschilds (and the British banking establishment more generally), and the Illuminati, a secret order dating back to Bavaria in the 1700s—a traditional bugbear of the political Right said to be in league with Freemasonry to control world political affairs. The meme has evolved over the years as a general reference to an imagined New World Order seeking to impose a despotic world government. The Tavistock Clinic played key roles in British Army psychiatry and development of psychological warfare. In ultra-right-wing folklore, it is said to have conjured up the Beatles and the Rolling Stones in order to demoralize Western societies.

"Wouldn't there have been a better way to disable Germany, or disarm it, than to completely destroy Western Europe, and Germany in particular?" I said, apropos of the Second World War.

"Destroying Western Europe was a very profitable thing. They made a lot of money at it. And they killed a lot of white people. What's not to like from their perspective?"

"Why would the British want to kill a lot of white people?"

---

33 Allen Welsh Dulles (1893–1969) was an American diplomat and lawyer who became the first civilian Director of Central Intelligence (DCI), and its longest-serving director to date.

34 A psychiatric clinic located in London. During and after the First World War, soldiers were treated for shell shock there. Not to be confused with its post-World War II incarnation as the Tavistock Institute for Human Relations.

"They're bad people, man," Rob said. "The Brits are real jerks. What they did to Germany after World War I, the Versailles thing, was just awful. How can you doubt British malevolence after Versailles?"

"Well, how can you doubt German malevolence after Auschwitz? How can you doubt Russian malevolence after the gulag?"

"Because those were forces that were set in motion by the British in the first place."

"How does that work?" I asked.

"Who funded Lenin, gave him twenty-five million dollars, which is about a billion dollars in today's money? Jacob Schiff, from the law firm of Kuhn & Loeb, in New York. The Russian Revolution was funded from New York City. And so he [Lenin] showed up in a sealed train with twenty-five million dollars to do the Bolshevik Revolution."

"What was in it for Jacob Schiff?"

"I translated two chapters of Solzhenitsyn's book from Russian into English," Rob said. "The book is called *Two Hundred Years Together*, and the period that I was translating was between 1914 and 1917, and Jews were very sympathetic to communism at the time because they believed it would reduce anti-Semitism. In the first half of the Russian revolution, the first law that was passed was against anti-Semitism and punishable by death."

In fact, Solzhenitsyn blamed "the renegades" among both Russians and Russian Jews for constructing the totalitarian Soviet state.

"Why was the Russian czarist government so weak that they could be overthrown by this gang of Bolshevik riffraff?"

"First of all, twenty-five million dollars helped," Rob said. "They got bribed. Just like the paycheck liberals have today—you just bribe the shit out of them. Twenty-five million dollars in 1917 is like the equivalent of a billion dollars, because two cents in 1913 is a dollar today. So they could rev up a color revolution. And they moved pretty quickly. They had the cadres ready to go, and once they had the funding it was just a matter of time. I mean if there was not that twenty-five million dollars from New York City there's no way what happened."

The twenty-five million dollars is apocryphal, of course. It seems to derive from, of all things, a *New York Journal* gossip column in February 1949 by the popular Cholly Knickerbocker (actually a pseudonym for a rotating set of reporters with Igor Cassini, brother of fashion designer Oleg Cassini, presiding at that time). In the column, Jacob Schiff's grandson, John, told Cassini that Jacob Schiff had given twenty million dollars to the Bolsheviks. It is a fact that Leon Trotsky was in New York for nearly three months in the winter of 1917. At that time, Jews made up a fifth of New York City's population of 5.5 million. Many of them had fled Russia to come to America. Many were sympathetic to the new socialist and Marxist movements that roiled Europe as a response to the nineteenth-century rise of industrial activities. Many of these worked in the sweatshops of New York and had become active in American labor politics and the rise of unions.

Trotsky (born Lev Davidovich Bronstein, a Jew) excited the large Jewish community of the city with his denunciations of the czar's anti-Semitism. He published articles in the Yiddish New York paper the *Jewish Daily Forward* (*Der Forverts*), though he was not raised speaking Yiddish and couldn't write in it. The paper had a circulation of 200,000. He gave speeches. And he swanned around the city with his family in a limousine provided by an unnamed wealthy admirer.

What was in it for the banker Jacob Schiff, a personification of "capitalism," to support revolution in Russia? There is scant record of his disposition. While Trotsky was in New York City, a February 1917 uprising in St. Petersburg against the provisional government of Alexander Kerensky led to the abdication of Czar Nicholas II. A rally celebrating the event was held at Carnegie Hall on March 23. Jacob Schiff did not attend, but the *New York Times* published a telegram sent by Schiff that had been read to the audience, saying he expressed regrets, that he could not attend, and then describing the successful Russian revolution as "what we had hoped and striven for these long years."[35] Perhaps Schiff simply considered the czar an enemy of the Jews and believed that Trotsky and his associates would put an end to anti-Semitism there. It is not known if Schiff understood that the Bolsheviks expected the revolution to spread all over Western Europe. In any case, Russia's disastrous involvement in the First World War had, by early 1917,

---

35 "Mayor Calls Pacifists Traitors," *New York Times*, March 24, 1917, p. 2.

shoved the nation into economic and political chaos, with military casualties of roughly five million dead, wounded, captured, or missing, leaving it weak and subject to insurrection. The Western world had little experience then with socialist governance, or with the despotism it would soon evolve into. In the years that followed, the Soviets liquidated millions of their own citizens, including many educated professionals and intellectuals who posed a threat to the Soviet narrative, and hence the regime's legitimacy. I asked Rob if he thought the Russian population had been dumbed down as a result.

"I don't think so," he said. "They're very intelligent people. They're really good at math. I mean, probably, but I think the whole world is dumbed down these days."

BRINGING IT ALL BACK HOME

Yet, here he was now: Rob Freeman, a middle-aged family man still dabbling, when he could find the time, in a young man's game of political agitation. He left Boston and came to the Quiet Corner of Connecticut in 2003 because he could afford to live there. His wife, Anna, was working as a licensed attorney and Rob worked for her for years as a paralegal. He hit a speed bump when Anna ran for an elected probate court judgeship and her opponent discovered Rob's alt-right connections—and "doxxed," or outed, him publicly. Anna lost the election.

He wanted to run a business of his own, and he liked the idea of a cab company.

"I wanted to be the mass transit operator of Windham County. But also I didn't want Muslims to start a taxi company here."

"Were they doing that?" I asked. "Here or elsewhere?"

"Yeah. Taxi companies are typically owned by foreign people. And mostly Indians open convenience stores and liquor stores. This is what the Thomas Friedman/Bret Stephens *New York Times* guys are like: *foreigners are so great, so wonderful, because they open businesses . . . aren't they wonderful!* Well, let me tell you something about the foreigners opening businesses here. They have somewhere to flee after they get caught stealing everything that ain't nailed down. They stiff their creditors, they stiff their employees. They stiff the IRS. They stiff the state. They stiff the local taxes. They don't pay anything and they just put the money in their pocket as much as possible, and then when they finally get caught, they go back to their home country. That's why foreigners like to start businesses here: because they get the money, and they don't pay anything."

"Surely some of them really want to stay here and play by the rules."

"But a lot of them are doing that model," he said, "and they find all kinds of ways to game the system, and because they're 'diversity,' the government lets them do it, the government looks the other way."

In 2014, Rob Freeman set up his cab company with $20,000 of his own savings plus $80,000 in a business loan. He continues to battle the intermediary company that is supposed to channel the state's payments of his invoices and always drags it out as long as possible. What you have read are his unadulterated ideas and beliefs about the current state of our national life. This is obviously not a comprehensive survey of alt-right ideas and activities across the USA, but of an individual with his own singular history. Many readers will find much of this material unappetizing, but at least you know what one particular self-described white nationalist is thinking. It's better than not knowing.

As with everyone else I talked to for this book, I asked him the same concluding question: *Now what . . . ?* He answered in an email some months after my visit to the Quiet Corner.

> Buddhism, I've found, is how to live in Hell without becoming a devil. The competent man who guards his thoughts and emotions and stays calm and studies differential equations and provides for those around him financially even while living among very unhappy and unhealthy people—that is how the Aristocracy of the Competent is formed. When the Competent beta males rise to the top, that is when we will have a Renaissance.

Chapter 7

MAKING WHISKEY IN THE HILLS

I've been drawn to people who are determined to carve an independent life for themselves at practical vocations. In particular, I'm interested in careers outside of the spiraling corporate and bureaucratic insanity that now holds many citizens hostage to mercenary livelihoods and that provokes so much anxiety, anomie, alienation, and despair. Quite a few of these independent souls have found me because of the books I've written about the prospect of an economic collapse that will dramatically change everyday life in this society. Around the time I conceived the plan for this book, I got a letter from a stranger named Kempton Randolf, who wrote:

> In 2001, I think it was, I went to a lecture you gave at Skidmore [college] on your Geography of Nowhere *book, where I was a biology undergrad at the time. I must say, that was probably the most pivotal presentation of my life. It resonated with me at just the right moment, and in an odd way it was quite inspiring . . . After graduating Skidmore, I moved to Cabot, Vermont, bought a run-down 200-year-old farmhouse and completed the Historic Preservation graduate program at UVM [the University of Vermont]. Not satisfied with the typical preservation occupations that mostly consist of wagging fingers at the owners and developers of historic resources, I sold my family home in Massachusetts and bought 60 acres of farmland surrounding my old farmhouse, and decided to make a go of it at real-life full-time farming in 2010. To make this brief, seven years later, I could not be happier . . .*

In further correspondence, I learned that Kempton had taught himself the art of distilling liquor and was one of the very few people in the USA who was making whiskey and other spirits commercially from grain he grew himself. So in early May 2018, I drove up to his Hooker Mountain Farm Distillery outside the tiny village of Cabot, at the edge of the region they call "the Northeast Kingdom." It was only twenty miles east of the state capital, Montpelier, a lively and sophisticated town full of gourmet bistros, brew pubs, cafés, a bookstore, and all the other trappings of hipsterdom, but twenty minutes beyond the city limits, it felt like the back of beyond. There was a whole lot of nothing up there besides woods, and I was faked out by the missing sign for Hooker Mountain Road.

"They changed it because everybody was stealing the road sign," Kempton would explain later. "It used to be Hookerville Road. They were spending too much money continually replacing the sign. So, I have a sign that I put out on Route 2 when we're open."

The catch was, he was only open on Sundays, and it was Wednesday. Kempton's sixty-five acres are actually on another dirt road off of Hooker Mountain Road called Lovely Road, and it was just that, though spring was lagging about two weeks up here at 1,400 feet above sea level. In New England that can make a big difference. Lovely Road ran alongside the shore of Molly's Pond, a reservoir with no houses along it, lending the vista a back-in-time look of the days before European settlement. The woods were "lovely, dark, and deep," as Robert Frost once put it, and pink buds on the trees would burst into leaf any day. After a mile or so, the forest opened up into fields and pastures, and the distillery came into view. It looked like a barn, but unlike most barns in the old northeast, the distillery was obviously new construction. The roof showed no sign of sagging. The board-and-batten siding was painted a sedate slate gray, set off and brightened up by cheerful red double doors and a large red sign above them with "HOOKER MTN. FARM DISTILLERY" spelled out in handsome capital letters complete with serifs. Off to the left was the old red-painted farmhouse. Beside the barn was a complex of sheds, pens, and additional barns for the livestock—pigs, goats, chickens, and three dairy cows.

Kempton expected me, of course, and he soon appeared from working out back among the animals. He was tall and lanky, with long brown hair tied back and a scruffy beard. His face was chiseled, Lincolnesque. He wore work boots, heavy-duty work pants held up with buff-colored suspenders, and several shirt layers up top. It was still nippy out. We didn't waste much time in small talk since he had set aside the day to talk to me, we had a lot of ground to cover, and it was the time of year when a thousand and one distracting chores beckoned.

At the center of the distillery was an enormous masonry stove, sometimes called a Russian stove because they are a fixture in traditional Russian houses. It consists of a firebox within a large masonry mass. The idea is that you fire it for a short period of time, and the masonry retains the heat for hours and hours. He'd hired an artisan mason to build the thing, which cost about $15,000 for the unit itself, and another $10,000 or so for the masonry deck that surrounded it at an upper level for managing the three-hundred-gallon steel pot still that he designed his business around. A separate room off the distilling parlor had the look of a frontier laboratory, the shelves and workbenches lined with bottles, jugs, flasks, beakers, retorts, funnels, condensers, and other accessories of the liquor-making trade.

Kempton pretty much learned the process on his own, by trial and error. He had to build the whole setup before he could even apply to the state of Vermont for a distillery permit.

"It's a very burdensome application process," he explained. "Lengthy and difficult. Most people hire a consultant to do all the paperwork for them. I did it all myself. It took about three months, and then it takes another four months after you submit it until they get back to you, and then you've got another month or two months of corrections and back and forth and stuff. So, it was a year just to get the permit. And then, once I had that, I started distilling. But I was left in the position of, okay, now you have a distillery, and you can distill—and I had never distilled before. I had never done it. I was doing some water distillation, steam distillation to make essential [herbal] oils. That's how I cut my teeth on all the equipment because that's not illegal. You can do that without a permit. The exact same equipment. I was doing hydrosols and essential oils to figure out how to use all this stuff, and then when I finally got my permits in place, then I just went from there."

He had to trust his nose and his palate to tell him what worked.

"I did a lot of research and reading and looked into a lot of historic methods of producing alcohol. My production methods are very different than any other distillery because I had to come up with them all myself. It's kind of like home distilling on steroids versus an industrial process—which is what happens at most distilleries."

His first commercial batch was a potion he called *Sap*, made from a mash of mixed corn and oats, aged on sugar maple wood and blended with actual maple sap, the precursor to syrup.

"I blend it with maple sap instead of water. The reason is because we have pretty specific laws in Vermont about what's an agricultural product and what's not. There have been court rulings where other distilleries that really are not farms were trying to take advantage of the relaxed regulations that farms are subject to in Vermont versus other businesses. The court said, if it's more than fifty percent water, it doesn't count as an agricultural product. I was very determined to have this be an agricultural processing facility. So, I said, 'Well, fine, then I'll just blend it with maple sap.' I freeze maple sap in the springtime fresh out of the tree, thaw it, and then use that to blend. It's a little alkaline. It has a higher pH than just water does, so it's pleasant to add but it doesn't impart any flavor."

Kempton added a little maple syrup to sweeten the liquor slightly.

Most whiskeys are less than 50 percent grain alcohol. The *proof* index stands for double the percentage of alcohol it contains. If a whiskey is 86 proof, the alcohol content is 43 percent. Anyway, going higher can affect taste adversely. And liquors over 100 proof—50 percent—can make your eyes water.

With any particular batch, he does two runs through the still. The first batch comes out between 20 and 30 percent alcohol. Then it goes back in the still and the second run comes out between 68 and 78 percent alcohol.

"When you're doing that spirits run, the second run, you're cleaning the alcohol. You're taking out the methanol [CH_2OH] and isopropyl alcohol [OH] and leaving ethanol [C_2H_5OH]. They all have different boiling points and some methanol comes out first. That's 'the heads.' And then you have the heart of the run, which is the ethanol portion, that's about 70 percent of the distillate that comes out. And then you have 'the tails,' which has most of the isopropyl alcohol. It's saved and gets put into the next batch. There's still some more ethanol in the tails, so the tails get recycled back through."

Methanol, otherwise known as wood alcohol, is a toxic substance that can blind people by destroying the optic nerve. A mere third of an ounce will do the job. The median lethal dose (i.e., it will kill you) is about 3.4 US fluid ounces—roughly the amount of liquid in a dry martini. Wood alcohol was a great hazard during Prohibition, when a lot of bootleg liquor was made carelessly.

With all this exacting chemistry, and even in the early going, he's had only a few batches that came out badly. One was made with a mash of slightly mildewed grain. The moldy off-taste came through the multiple distillations. He was reading up and studying constantly. He got a little instruction from an old-time local distiller named Duncan Holliday, who had an operation called Dunc's Mill over in Barnet, VT. Dunc was retiring and, in fact, sold his outfit the year before I came to Kempton's place. Dunc also taught a distillery course every spring at Vermont Technical College, though Kempton did not enroll in it. By coincidence, the class was coming to see Kempton's outfit the following day.

There are about fifteen distilleries in Vermont these days, he told me.

"I'm friendly with a few places," he said. "More of the small-scale kind of agriculturally focused distilleries. There's a pretty big divide in the distilling community, actually a few divides. One divide is on the scale of operation. Some places are getting very large. The other is on process. A lot of places that call themselves distilleries don't distill anything. They buy bulk alcohol and they stick it in a barrel, and then they stick it in a bottle, and they put their name on it, and they sell it as their own. Basically they're just aging somebody else's alcohol. If they're even doing that much, because you can buy completely finished ten-year-old whiskey that comes in a bulk container, drain it right into—"

"People do that?" I said, naively perhaps.

He explained that maybe seven of the fifteen do distill their own product, but at the time we talked, Kempton's was the only Vermont outfit that distilled using its own grain grown on-site.

"There's one other distillery in the country that I know of that's growing their own grain really. They're in Virginia and they've been operational since the 1930s. They have a large farm and they've been growing their own grain for a very long time."

Kempton started growing his own grain before he conceived of operating the distillery. It was after he first settled into the farm, and he grew oats, barley, and flint corn to feed the animals he was acquiring.

"There are several large macro categories of types of corn," he explained. "Flint corn, dent corn, flour corns. There are oil corns. They have a high oil content. Flint corn is a high-protein corn. It's a very hard outer layer to the kernel. The vast majority of corn grown in the US is dent corn. Flint corn was given to the Vermont settlers by the Abenaki Indians, in the late 1700s. The settlers improved it. It was found growing on a farm in Calais, Vermont, about fifteen years ago. This guy had the strain from his grandfather, and High Mowing Seeds [a commercial organic grower based in Wolcott, Vermont] bought it from him and released it. I bought it from High Mowing Seeds. They turned it into a commercial variety that you can buy. It's open-pollinated corn so you'd save the seed every year, and it's the shortest season corn that's in existence: seventy-five-day corn—"

Just then, a six-year-old boy came into the barn. This was Avi, the son of Kempton's wife, Carrie, from a previous marriage. He was three when Kempton and Carrie found each other via the OkCupid online dating service in 2014. They've had two other children since then: Mallow, who was two and a half at the time I visited, and Cricket, just six months old. Avi had been sent to the distilling barn to fetch us to the house for lunch.

THE FARM

Inside, the house was the kind of semi-chaotic scene you'd expect with two infants and a six-year-old at large. We settled at the table in the warm kitchen and ate a bountiful meal of roast pork raised on the premises, sides, and plenty of warm homemade bread with fresh butter. Carrie was in charge of the dairy operation: the three cows. She was working as a Montessori schoolteacher when she and Kempton came together. She was from a town way up north, Derby, a few miles from the Canadian border, but she had not grown up as a farm girl. Her parents were not native New Englanders either (Dad, Maryland; Mom, Idaho). Her father worked for the US Geological Survey (USGS) as a geologist and soil scientist.

Kempton had mentioned earlier that he'd had several different girl-friends before Carrie came along.

"The problem was, nobody wanted to really do this," he said. "Nobody cared. It's hard. Some of them kind of thought they wanted to live that life. But then when it actually came down to it, they didn't want to do it. And so being here and living this kind of life, it's always been kind of the sticking point where it didn't really work out. Then I found in Carrie somebody that I got along with really well and who equally wanted this kind of life."

When we were finished eating, Kempton took me on a tour around the spread, visiting the animals in their various quarters, and then out to a new field he was preparing for cultivation. It was full of stumps and tree debris. When he acquired the property, it hadn't been farmed for decades. Of the sixty-five acres he started with, only two acres were fields where grains could be grown. He's added a new cleared acre a year. A neighbor helps. ("He'll pop all the stumps up and then I still have a lot of cleanup work that I have to do with the tractor.") It's very hard work, and now he's up to six tillable acres for grains and potatoes, though he also clear-cut fifteen additional acres to use as pastures for the grazing animals. They'll keep the brush down. The soil is loamy glacial till, a gift of the retreating glaciers.

"It's actually pretty good," he said. "It's very well drained. There's been so much work done here over hundreds of years, over generations, and one thing that people don't realize: if you have a field in Vermont around here, and it doesn't have any rocks in it, it's because somebody spent so much time taking all the rocks out. There's such incredible investment from generations that are gone. I have fields that were cultivated that are rock free and that's because somebody went through and picked out all the goddamn rocks over a long period of time. That's a huge thing, and that's something that people don't think about, especially when we just let some fields grow back into forest and all that effort and the value that that cleared land had—you're turning it back over to woods and letting it slip away. We have investments from all these generations that we just totally abandoned. So, I have some fields that are fairly clean.

"I have the agricultural census records for this property from 1850, 1860," he continued. "I might have 1870. Have you ever looked at the ag census records? They're very, very detailed. They go through commodity by commodity, how many livestock, what livestock the operation had, what they grew, how many acres they had in cultivation, how many bushels of all the different ag products they produced. I found that ag census stuff in 2005 when I was researching the house and the property. I went back and looked at it last year and I was shocked at how close what I'm doing now is to what was done here in 1850. I mean I'm growing the same crops they grew: oats, potatoes, and corn. I do some squash and a couple of other little things, but those are my dominant crops. They had four dairy cows. I have three. They made, I think, five hundred pounds of butter a year off their cows, made so much in hay. They actually made a little bit more grain than I grow now, and that was with two oxen and a horse. And it's just shocking to me how similar they are. And it wasn't like I had that as a recipe and I said that's what I'm going to do: *I'm going to make this farm like it was.* I was just farming in a holistic, very traditional manner, and that's what works here. It turns out it's the same thing that people one hundred and fifty years ago found that worked."

He doesn't grow any wheat. It doesn't do well in New England, and it hasn't from way back in colonial days when a persistent disease called stem rust stowed away on ships from the Old World and established itself here. Barley didn't work out either, at least so far, though it is the staple grain of whiskeys made in Scotland.

"It's too cool and it gets fairly stunted and short," he said. "I have a combine that I use to harvest the grain. It doesn't do very well on barley. The oats run through it great. The barley, when it's grown in the field, if you just touch one of the heads they start to fall off. And because it's short, the combine doesn't cut it very well and there are just issues with it. So, I picked something more reliable."

He uses almost no synthetic fertilizers, herbicides, or pesticides.

"The only thing I've used like that is because I've had problems with ravens eating my corn seed. There's this powder that you can take and you dust all your corn seed in before you plant it. It's not toxic. It doesn't kill the birds but it irritates their digestive tract enough that it leaves such an impression on them that they leave your field alone. But it's not certified for organic use. That's the only thing like that that I've used. I had a whole field of corn that didn't grow one year because so many of the ravens came in and ate it."

I made a crack about Kempton's life looking like one never-ending science project, and that brought him back to his college days at Skidmore. One year, when he was living in a student apartment complex on campus, he tapped ten maple trees and lugged the sap back to his place, where he boiled it on the electric stove until water ran down the walls from condensation. He also installed the first solar electric panel on campus without permission from the school authorities. He fastened the panel to the roof, routed wiring over the edge, and ran the lights in his room from it. The school threatened to take it down. Kempton happened to be editor of the college paper, the *Skidmore News*.

"We said, 'Well, if you want to take it down, then the newspaper is going to write an article about how you guys just took down the first solar panel that was ever installed on campus.' And they said, 'Fine, you can leave it up.'"

A RUGGED CAREER PATH

Compared to the rough and hardy life he's living now, Kempton's origins were pretty genteel. He passed his early childhood in Andover, north of Boston. His parents divorced when he was three years old—though he says they remained very close afterward—and he lived in an apartment with his mom until they moved to a house in Newburyport when he was thirteen. He has no siblings. His father was a Xerox copier technician for thirty years.

"I get most of my love of old buildings and love of traditional skills and craftsmanship from him. He was trained as an airplane technician, and is quite a capable shade-tree mechanic and woodworker, who's particularly fond of diesel Volkswagens."

His father is retired now and visits the farm and his grandchildren about once a month.

His mom was "an electronics geek." She started soldering printed circuit boards for hourly wages and eventually a built a company of her own with thirty-five employees.

"It was a completely woman-owned business," he recalled. "And they employed mostly women. They were good at the work, very fine motor-skill type of work. So that was very formative for me. I had some idea that I would work for myself because she ran her own business."

His mother died of cancer at fifty, when he was a sophomore in college.

Kempton majored in biology at Skidmore as well as editing the college paper. After bailing on the master's program in historic preservation at UVM, he was in a quandary about where to turn and defaulted to what he knew, the newspaper business, landing a job on the *Montpelier Times Argus* where he did graphic design, layout, and some editing. He was living with a girlfriend in Cabot, a half hour away. The property on Lovely Road came to his attention around that time, and he researched it thoroughly before buying the house and eight acres in 2005. He became increasingly aware that the *Times Argus* was "a sinking ship," as was the case for virtually all local newspapers around the country once the internet got traction.

Meanwhile, he began to renovate the old farmhouse with money that he got from selling his mother's house back in Massachusetts. One of the first things he did was rip out the ceiling above the living room. Hundreds of corncobs fell out.

"It was the mice," he said. "They just filled the ceiling with corncobs. A hundred and fifty years of mice will move a lot of stuff."

Eventually, the man who sold him the place agreed to sell Kempton an additional fifty-five acres adjoining. He began actively homesteading while renovating the house, with a couple of chicken coops and some piglets.

Also meanwhile, his then girlfriend up and moved to Ireland.

"I did a lot of homesteading stuff for myself for a few years, just mostly getting familiar with livestock and carpentry and working on my house here. And then when I decided in 2010 to start farming, everybody told me I was insane. I think I had read too much Joel Salatin.[36] I started raising meat birds and pigs on pasture. I vividly remember going to pick up the piglets. The farmer stuck them in a feed bag and handed them to me. He said, 'Just put them in the foot well of the car and hold the bag shut, and they should be fine.' So, I started doing that in the backyard and I really enjoyed the process of procuring my own food. I did the slaughter and butchered pigs, did all the cutting, did some smoking and curing of the bacons. Then I got some beef cows and started doing that, and for a couple of years just did meat."

"Who were you selling to?"

"I was selling to locals, mostly by word of mouth. I didn't have any products in any stores. I was going to a couple of very small farmers' markets. Sales were pretty slow. It's hard to get people excited about chicken. When you're producing a commodity like that, you're under such pressure from what you can get at the supermarket, even though it's not the same product. There's still a big price pressure, and grain is very expensive here in New England compared to the rest of the country because it's very far to transport it up here. And so, yeah, the economics weren't very good."

By then, Kempton had ditched his job at the newspaper and was working at the farm full-time. By and by, an old college buddy named Dave came aboard to work with him, and he was there for three years. Kempton was paying him wages. Eventually, they had a falling-out and are no longer friends.

"We didn't have an operating agreement together. And it's hard to get money out of a business that doesn't make any money. So, that didn't work out well for him, compared to what his expectations were upon leaving. I had another guy whose idea it was to start the distillery and who was really interested in distilling. I'd met him through some other friends. He helped me get the distillery off the ground and the plan was that he was going to be the distiller here. And then he left before it got going. He quit in a big huff."

---

36 Joel Salatin: American farmer, lecturer, and author whose books include *Folks, This Ain't Normal*; *You Can Farm*; and *Salad Bar Beef*.

Then Carrie came into the picture.

"That was the end of 2015 and the distillery wasn't done. I had no distilling permit, so it wasn't operational yet. We still had a lot of work to do."

"During this time when you're in the learning phase," I said, "did you have some notion that you were going to get serious about doing it as a commercial venture or was it at this point kind of a 'science project'?"

"I made that decision when I started building the distillery in 2012."

While he went through the arduous process of getting that together, he made soda pop, which proved surprisingly to be a successful venture. The first product was a maple soda. He began using some of the other flavorings that he had learned how to distill. He got wintergreen flavor from distilling the twigs of the yellow birch tree. He added spruce, and lime, and orange-cream flavors (though the citrus fruit was obviously not grown on the property).

"It started as an accident. That guy Dave I had working here, we were in the sugar house. He had one of those SodaStreams, a little carbonation thing. And we just started mixing syrup and seltzer together and thought it was really good. I made a manual bottling machine. We started making it and selling at one of the markets, and people started buying it. And then that kind of took off. At the peak of that I had about fifteen different retail accounts, probably moving thirty or forty cases of soda a month. I did well enough to kind of limp along for a few years. And it took off so much that it did get in the way of starting the distillery."

We were back in the distilling house. He took a short, stout bottle out of a cupboard and poured a taste.

"This is a gin with spruce and a bunch of garden herbs," he said.

I toasted him. It was an unusual and provocative flavor.

"I've always felt that the future of farming is in value-added products," he said, meaning more process-intensive items made on the premises from things grown on the premises, multiplying the value of what you produce. "It allows you to farm at a smaller scale than being a commodity farmer."

One product he's excited about is a maple cream liquor made with cream from his own cows, maple syrup from his sugar bush, and egg yolks from his chickens. The alcohol acts as a preservative in the concoction.

"We stumbled upon this weird loophole in the dairy industry regulations where we don't have to operate a creamery because we're not making a dairy product. We're making a distilled spirits product."

"Meaning you're off the hook for pasteurization issues?"

"Yes."

He does sell some raw milk to a few select customers, and with three children, they use a lot on the farm.

"Basically, what happened during the twentieth century is we had a conspiracy to criminalize small farming and criminalize self-sufficiency and productive lifestyles," he said. "So, when you get into, oh, I want to make cheese and sell it to somebody, I want to make butter and sell it to somebody, you run into hurdle after hurdle to do that in a cost-effective manner. I have enough regulations and crap that I deal with trying to make my distilled spirits. With every product there's some hurdle that's been intentionally put there to keep people either doing something at subsistence level, or make them invest so much money that then they can't compete with the already-established businesses in that sector. Vermont has fairly decent laws. If you're at the volume we're at, and as long as people come here to get it, you don't need a license for raw milk. You have to have your cows tested for certain diseases, and post the results, which we've done in the milking parlor, but it's fairly permissive when you're at a very small scale."

At the time we were talking, Kempton said that his whiskeys and spirits made up about 40 to 50 percent of his annual production. The cream liquors were about 20 percent. And he had been working on a kind of French-style cider—pommel or *pommeau*—which is a blend of two-thirds cider and one-third brandy, a hearty 20 percent alcohol. Or 40 proof. The whiskeys are 45 percent, or 90 proof.

It was getting late, and I sensed that Kempton had a long list of other things to do before supper time.

"So that's our production," he said conclusively. "I've moved everything I can make. So that's good. Now we're just working on being able to make a little bit more product. It's been great for the farm to have the distillery here. It's the first time that I've been able to get people to come here. And people are excited about coming out here. I'm only open on Sundays. I can't keep stopping in the middle of everything. I would say last year probably a third of my revenue came from just sales out of here, which is good, you know, and the nice thing is that it drives the sales of all the other products that I have. I still make a maple soda in the summertime just for here. Things are kind of bare right now because we're in the springtime but I do vinegars, always have a bunch of vinegar products. I have meat in here. And those really move."

By a supernatural coincidence, just as we were wrapping up, two guys drove up to the little parking area in front of the distillery. They were middle-aged, prosperous-looking gentlemen in mail-order casuals. I pegged them as fly fishermen up from Boston or New York. They wanted to buy some whiskey. Kempton said he was only open on Sundays. They asked him to bend the rules. He resisted. They pleaded. He stood firm, and eventually the pair shook their heads in incomprehension and climbed back into their car.

"Jeez, what'd you do that for?" I asked Kempton. "You could have made a quick hundred bucks."

He said it didn't bother him. He'd made $1,200 at the Burlington Farmers Market the previous Saturday.

Outside the distillery, in the afternoon sunshine, the air was now mild and springlike. I asked him my inevitable question: *Now what . . . ?* How did he see the future of our national life shaping up?

"I don't think it's very good for most people. I'm a millennial. I'm unfortunately lumped in that generation although I'm kind of in this weird in-between phase. I was born in '81. The people born between 1980 and 1984 are the really youngest around who grew up in an analog world, in a world that made more sense than today's world does to people. I am part of the most useless generation that's ever existed in human history. Millennials know how to do less for themselves than anyone that's ever existed on the planet Earth. Home Depot and Lowe's, a couple of years ago they realized 'our business model is screwed' because this whole generation of people that's coming up, they don't even know how to use a tape measure. Never mind, like, I'm going to get some tile and go re-tile my bathroom. Forget that. They don't know how to use a handsaw or a tape measure or anything like that. I know people that are my age that don't know how to cook an egg. You know, if you don't know how to cook an egg, in my book you're not a fully formed adult. I don't think it's good. I think we're headed to a place where people are going to have to do a lot more for themselves than they do now. And most people who are under the age of forty don't have a clue about where to even begin with that kind of stuff. We have been completely separated from any life of productivity, any knowledge of how to make anything or do anything.

"Everything's left to specialists. There's this weird both overconfidence that people have—this misplaced self-confidence—but then also a real lack of self-esteem. So many people I know are just always amazed that, like, how did you learn how to do that? How did you figure that out? How do you do that thing? It's called being a human being. That's what we do. That's what we were made to do as a species: to have a problem and use our hands and our brain to figure it out. And that's been beaten out of people. I don't think we're headed into a very good place."

Kempton's chores were calling him away again, each one requiring a set of specialized knowledge and a range of experience to carry out, all of it hard-won. I headed back down Lovely Road toward Montpelier and Interstate Highway 89, and all the less than lovely furnishings of the wounded American landscape that Kempton Randolf had left behind.

Chapter 8

FIGHTING FOR LIFE IN SMALL BUSINESS

In these days of American economic insecurity, social discontent, political distemper, and gender jockeying, there's an awful lot of idle chatter about "strong women," but Suzanne Slomin is the real deal. She overcame a lot of adversity on a long, wayward journey to establishing her Green Rabbit Bakery outside the tiny village of Waitsfield, Vermont, and at age fifty she has found her groove.

The road from the Champlain Valley takes you up and over the jagged spine of the Green Mountains. It's springtime at the start, and winter still lags in snowy patches up in the elevations, and then it's springtime again when you get down into the Mad River Valley on the other side of the range. The ski season was over, and Waitsfield was sleeping it off. The town got thrashed pretty badly when Tropical Storm Irene ripped through in August 2011, and the Mad River overflowed, damaging the American Flatbread company; afterward, hundreds of local volunteers gathered to clean up the place where seven feet of water rushed up to the level of the pizza ovens. Several other businesses were smashed, including the local laundromat.

I continued through the old part of the village—a few remaining pre-automobile shop fronts preceded by an array of the usual American highway jive of drive-in this-and-that—and across the repaired bridge over the now docile river, and a mile or so farther onto an unpaved road until I found Suzanne's place in a sheltered bowl of cleared forest below Mt. Alice. Near the rim of the bowl stood the Green Rabbit bakehouse bermed into the slope. It was all new construction, designed with her friend, Danny

Sagan, head of the architecture department at nearby Norwich University. The unpainted wood siding was still bright and buttery. The bakery ran off a solar electric system based on three freestanding arrays of panels between the bakehouse, the garden, and an orchard of mixed fruit. A few sheep grazed in the vicinity of the solar panels. They were sweet and picturesque, though fated to end up as food. Suzanne's home, a renovated old farmhouse, stood farther along the rim. She was preparing to build a shed garage for the tractor and other equipment, and a bed of crushed stone was freshly laid onto the construction pad, awaiting the building crew. Below that, going down the side of the bowl, were the gardens and rows of small fruit (blueberries, raspberries, currants) where she grew many of the ingredients that would come out of the bakery.

Suzanne was expecting me. We hadn't met before, though we corresponded over the years about the condition of the economy and related issues. Suzanne was among those with worries about the future of industrial civilization, and that had a lot to do with the choices she'd made in life. But now she was intensely focused on running the business she had fought so hard to establish. She was petite, with short brown hair, rather pretty, actually (if it's okay to say that these days), and quite visibly fit and strong—no surprise, considering the amount of sheer physical labor she's put out for years. After a cup of tea and some chitchat, she took me on an orientation tour of the place.

Inside, the ground floor of the bakehouse was the heart of the operation, a big, sunny, open room with a high ceiling filled with sturdy worktables, racks of baking pans, sinks and workstations for washing hands and sanitizing utensils, plenty of cooler space for storage, and finally the triple-deck German bake ovens that signified the culmination of a long, tearful story of major equipment failure that we'll get to presently. The sense of order verged on the fanatical. Altogether, it looked like a very cheerful place to do a lot of hard and rewarding work.

To understand her technical commitment to the craft of fine baking, we must begin with Suzanne's main product: slow-fermented sourdough breads based on cultures of wild yeasts. By a strange coincidence, her business and her growing skill at bread making came together in the same years that US nutrition culture and the alt.medicine scene launched a multifront campaign against carbohydrates, bread in particular, wheat especially, and most specifically gluten, the elastic protein that makes bread chewy. Some of this came from promoters of the paleo diet, which consigned all carbs to a kind of purdah—it became uncool to visit the center aisles of every supermarket, where the factory loaves of "balloon" bread, the kiddie cereals, snack cakes, chips, cookies, and all sorts of "creative" Frankenfoods dwell in ignominy.

Some of it came from the naturopaths and the "integrative medicine" sector who try sincerely (for the most part) to unravel the mystery of the country's poor health—the range of bafflingly intractable ailments that afflict the American public including chronic fatigue, arthritis, dementia, obesity, diabetes, irritable bowel syndrome, "leaky gut," myofascial pain, and neurological conditions like multiple sclerosis and ALS (Lou Gehrig's disease). The message went forth: *get rid of the gluten in your diet!* No more sandwiches, no more toast and honey, no more milk and cookies. Pretty grim. Dr. William Davis opened up another front with his 2014 book *Wheat Belly*, which made the case that the commercial wheats hybridized since the 1970s—engineered to shorten the stalks and enlarge the seed heads—have resulted in grain products carrying a toxic load of Frankenproteins that the human body cannot process effectively.[37] Davis also made a case that one especially sinister protein, gliadin, messed with morphine receptors in the brain to make bread addictive.

Suzanne was well aware of what she was up against. Her pro-bread case comes down to the idea that good organic grains, when allowed to ferment properly, can be digested successfully by most people, and that five thousand or so years of human experience suggests that bread can still be a healthful staple food. Apparently many customers in central Vermont agree, because her sales are robust, even in a state with a large and diverse corps of alt.medicine practitioners militating against bread. It is also the case that her breads are delicious: the savory levain loaves, the olive-rosemary loaf, the pain de mais (wheat and rye with cornmeal), and so on.

Unlike the factory breads that can go from dry flour to a finished, packaged loaf in two hours, Suzanne's bread is "management-intensive"

---

37 William Davis, MD. *Wheat Belly*. Rodale, 2014.

and takes twenty-four hours to complete. At the heart of the process is her sourdough starter, a cultured slurry that is a kind of living organism. It has to be fed twice a day, like a cat. The wild yeasts consume the flour, reproduce, and give off alcohol as a waste by-product. They require a temperature-controlled cozy environment to thrive—though there are times when she refrigerates it to slow down the process—and Suzanne does whatever is necessary to coddle them. She sources her organic grains mostly from the Midwest; wheat is hard to grow in New England. During one particular year of freaky weather, she had to get wheat from all the way down in Argentina to keep up her production of two hundred to three hundred loaves a day, depending on the season. Once the loaves are formed and shaped, the fermentation process continues for twenty-four hours. It is hard and exacting work. The loaves wholesale for four dollars and generally retail for five dollars or more. Whatever bread doesn't sell, she buys back from the markets and shops who sell it, and she donates the loaves to churches and community centers.

She had two employees when I interviewed her in the spring of 2018: Clara, who assists with the baking, and Cam, a recovering addict who "was fragile at first and now is an integral part of the team." He makes deliveries and puts more miles on Suzanne's car than she does. Having been treated pretty harshly as an employee herself, she is very concerned with treating her own workers decently. The rule of thumb for her business is that employees should cost between 10 and 30 percent of the gross revenue. For Suzanne, it was closer to 40 percent. She could not afford to furnish health insurance for them, but she agreed to cover 50 percent of the deductibles out of their own health insurance if they got sick. She made a deal with Clara to pay her 60 percent of her salary for six weeks of maternity leave if she stays with Green Rabbit for two years and 80 percent after five years. Otherwise, she would keep losing employees to Ben & Jerry's ice cream up the road in Waterbury, which offers health coverage. Ben & Jerry's cultivates the aura of two hippie dudes running a boutique business out of an old Vermont barn, but in fact it was bought up by the global Unilever Corporation in the year 2000.

"America treats small business very badly," Suzanne remarked.

FROM ROOTS AND ROOT CELLARS

Suzanne was born on Halloween in 1969, raised on Long Island, and went to public high school in Rockville Centre there. For college, she went to Colgate University in a rural pocket of central New York State, south of Utica, which is everything Long Island is not: peaceful, bucolic, unchanging. After graduating with a major in philosophy of religion and a minor in studio art, she moved to New York City to construct a career. She is very deliberate about the moves she has made in life. She had good graphic design skills and tried to break into the vocation, but the work was mostly freelance, and the irregular pay and the never-ending hustle for work wore her down. Since childhood, though, she'd maintained an interest in cooking and baking, and she was beginning to think of it as a possible line of work. She waited on tables and worked early bakery shifts, and she pounded the pavements with her art portfolio in between. By and by, she got a pretty cool job installing and maintaining exhibitions at the American Museum of Natural History, but she couldn't shake the nagging desire to enter some part of the food industry, so she enrolled in a six-month course at the French Culinary Institute.[38] For several years after that, she worked in restaurants and catering companies in Chelsea and the East Village. The city was wearing her out, and in 1999, her thirtieth year, she split for a cooking job in the Berkshire Mountains of western Massachusetts.

"I was cooking at a very high-end inn called Blantyre, which was like a Scottish castle, and it was beautiful," she recalled. "But that was the first time I'd worked in real high-end dining and it just seemed stupid to me, like really dumb. I realized at that point that I didn't really have a path in cooking because that's what you were supposed to move towards. I had always been interested more about where your food came from and for years had wanted to learn more about agricultural production. The nice thing about working in that place is that they sourced a lot of their foods locally when they could."

---

38 At the time, the French Culinary Institute was located on Broadway and Grand Street in Soho. In 2011, it moved to Silicon Valley and changed its name to the International Culinary Center.

The guy who grew the salad greens that Blantyre served was Ted Dobson, known as "the Godfather of Mesclun," the dainty baby lettuce mixes developed in California that he helped popularize in the East. Suzanne got to know him and, after leaving Blantyre, started doing seasonal labor at his farm in nearby Housatonic, Massachusetts. Through the winter she worked in a retail wine-and-cheese store. Dobson had a keen interest in changing the model of American farming. He was convinced that industrial agri-biz was ruinous and doomed—"[the] genetically altered, petro-slurry-fertilized corn grown year in and year out on the same land to feed herds of bloated dairy cows," he wrote on his blog—and Suzanne was inspired by what she was learning from him. She also met Aaron there, the man she would marry.

Dobson's farm was under pressure from developers, and he would eventually shift his operations to Sheffield, Massachusetts. After finishing a second season with him, Suzanne and Aaron signed on to work on an established organic farm down in Pennsylvania. They were thinking about getting a farm of their own and wanted more practical experience. It only lasted about a month.

"The fellow [the organic farm owner] was a bit misogynistic and he kind of picked on me," she said. "I decided I wasn't going to stay. And then we both ended up leaving, just went back to Massachusetts to work for the summer and decided we would look for land, and then realized that most places there were pretty unaffordable."

She had loving memories of the area around where she went to college at Colgate, back in central New York, and they began scouring the region for a farm. They finally found one in the rural township of Madison, five miles from Colgate.

"It took a lot to find a piece of land. We were getting soil maps and driving around, and we did meet with some farm realtors but ultimately we ended up just asking a dairy farmer. There was an obvious piece of bottomland and a barn that wasn't being used, and it was owned by the adjacent dairy farmer. We asked one of the farm brokers to just ask if he'd sell it, and he said 'yeah.' He had corn planted there for the year. But we realized that since we knew we would want to be certified organic, we actually needed to buy an additional five acres to have the buffer from his crops. That actually added an additional seven or eight thousand dollars. After that was said and done, we had about forty-five acres and a barn. We continued renting a little house in the area, and then we built an apartment into the barn, where we lived for the first few years."

That proved to be problematic. They learned that they couldn't get insurance for the barn if they put a woodstove in the apartment inside it.

"At first we just didn't install any heating. We just put in an oven and spent the first winter constantly cooking and baking. And then we ended up putting in a little propane heater."

Eventually, they decided to build a passive solar house on the property. That was the first time Suzanne encountered Danny Sagan, the architect. She'd designed the house herself, but needed to consult a licensed architect to make sure the building would actually work.

After the frame went up, they finished a lot of the interior themselves.

"And then we started out just farming, mixed vegetable farming, and we had maybe a dozen laying hens."

About a year later, Suzanne heard "through the grapevine" that a guy in the vicinity was selling a bunch of baking equipment: a sink, mixers, workbenches, and ovens. He'd been running a hobby bakery in his basement but was stricken with multiple sclerosis and couldn't manage it anymore. She and Aaron had built root cellars in the barn, and they converted one into a bakery.

"It was a horrible bakery," she said. "The ceilings were super low. We cut a big hole in the upstairs of the barn floor above the pizza ovens, and put in a big barn fan, and that was our ventilation system."

"Sounds like a Navajo sweat lodge," I said.

"It was. In the summer mold would grow on everything and we'd have to constantly scrub things down, and then in the winter we'd have to heat it to keep it warm enough for the bread. And the only way to really do that was electric baseboard, and so that dried everything out. It was fine to get started, it was a horrible facility."

By their second season there, they were able to go to the local farmers' market with bread and vegetables.

"I remember the first thousand-dollar week we had. We were barely underway and I was like, wow, you can actually make this work! We sold a lot of bread and only baked twice a week, but by the second year of doing the same farmers' market we would sometimes do two thousand dollars cash—at just one little town farmers' market. It was pretty shocking. By the third year there we were at the forefront of locally grown. We had a little farm stand, we had restaurant accounts, and we had some retail shops. Syracuse was about as far as we went. And we did a lot of business."

Back then, they had some time off in the winter.

"Thanksgiving you could really celebrate and feel good, if you had a good year, and then take a breather. Which I actually didn't do that well with. I need a constant regimen."

At that point, Suzanne baked once a week. She delivered to a health food store in nearby Clinton, NY, and had at least one restaurant taking her loaves. Altogether, they ran the farm-and-bakery operation—which she had named Green Rabbit, from a dream she had as a child—for six years. By that time, they ran into new problems. Suzanne's mother-in-law got seriously ill. She and Aaron were having trouble getting along as a couple—and hadn't married yet. She wanted to open a farm store, something more than a seasonal stand.

"Aaron really didn't want to do that and it became a big bone of contention. I wanted more of a year-round structure, and he really didn't. At one point we were talking about one of us getting an off-farm job. I remember we were hoping that Aaron might get a job with a very successful and well-known local blacksmith, but he ended up not getting that job. I looked at things like the cooperative extension, and there really wasn't any other work. The solution was maybe to sell [the farm], and get out of that business, and out of working together. The goal was to relocate and get jobs—not continue farming or anything like that."

They talked about finding something in New England. Aaron's ailing mother was in Connecticut. They traveled all over the region, and when they came to Vermont's Mad River Valley, the place seemed perfect. Among other things, the local food movement was a much bigger deal in Vermont than back in central upstate New York. The present location of Green Rabbit was only the second piece of property they looked at.

"We'd seen this place on the internet and I'd been communicating with a realtor, and we hadn't even sold our place in New York State yet. This was all kind of happening at the same time. And my criterion was I didn't want to move somewhere unless I had friends, family, or a job in place. I had actually applied to a business called American Flatbread, which still exists but it's very different. It seemed on paper like it might be the perfect fit for me. They had gardens, they had a big pizza oven. They make pizza or, as they call it, 'flatbread.' They said it was, you know, all traditional processes, which it isn't. It's just what we call *pizza-with-propaganda*. They're still here. They did have several locations. Now they only own the one here in the valley. They also had an entire frozen pizza line, which they've sold off to a company in New Hampshire. It was a pretty functionally run business. So, I had a job in place before we actually lived here."

American Flatbread was primarily a restaurant. Suzanne was doing some part-time cooking there, but it wasn't enough work.

"I did day prep and I didn't really work nights. I was kind of done with that. I was approaching forty. For that stretch of time I was working my four days a week here and going back to help pack up the farm while we tried to sell it. Aaron stayed there. We just had a few different short-term rentals during that process. It was incredibly frantic. Then we had buyers and that worked out. There was this young couple that was given an early inheritance, and they literally bought the place flat out. We did well and were able to move here. We made an offer on this place that wasn't accepted and then ended up renting down the road, just coincidentally. And about a year later, the owner of this place had the realtor contact us and say she was ready to drop the price because the septic was fouled, and I knew the place was a wreck. It actually worked out very well since we had a rental down the street. We stayed there and spent about six months renovating the house to the point where we could more comfortably move in."

It was still uninsulated, and they had yet to replace the ancient, leaky windows, which they did the following summer. In the summer of 2008, Suzanne and Aaron got married and then embarked on yet another Great Vocational Misadventure. The employment situation proved haphazard.

"I worked at a few different restaurants," she said. "This is pretty common when people move here. You just job hop. I worked at American Flatbread. They turned out to be a little dysfunctional and not enough hours, so I took a second job at another little restaurant in town. I was working at both for a while. Then at that place, the owner was crazy. I did end up working for him full-time for a while. Then American Flatbread lost their chef before the next busy summer season, and they asked if I could come back and kind of run it. And I assumed that after that season they'd put me on salary. At the end of the summer they told me I could apply for the job. [Laughs.] So, I left there and then I spent a few months not working. Then I worked for somebody doing landscaping, and this whole time my husband [Aaron], who was dead set on not getting back into farming, was kind of insisting on getting back into farming with taking over the farm that the food bank was running. I went into that begrudgingly. I didn't want to, but I also was slowly realizing that there wasn't a lot of work around here that made sense."

The Vermont Foodbank is headquartered in Barre, near Montpelier, the state capital. The food bank had bought a twenty-seven-acre property along the Mad River down in Warren, the next town south of Waitsfield on state route 100. Only about twelve acres were actually suited for growing crops. The soil was sandy. There was an abandoned gas station on it, along with the parking pad, and an old farmhouse, which needed a lot of lead paint and pipes removed if it was going to be rented out. The Vermont Land Trust had previously bought up the development rights, so the property could only be used for farming. In any case, it lay in the floodplain of the river, which would prove problematic.

"A lot of this happened around the time you turned forty," I said. "Were you having an existential crisis about, like: *Where am I? What am I doing?* Or were you too busy, or what?"

"I think I was disappointed," she said. "My personal life wasn't working that well. I felt like I had really torn myself away from home [central New York]. I felt like that was home. So that was hard. And we had put so much into it, even though it wasn't working. We were fighting a lot. I never thought that I'd build a house. I never thought that I'd be part of any of that. It seemed pretty huge, and then we just left it. I know there were some good solid reasons to do that. And then it seemed when we got here, I think I was just confused. We'd made some personal decisions that we wouldn't work together, we'd just have jobs, we'd have a life, you know, and then all of a sudden we're going to be working together and running another farm! I had a lot of reservations about that. I went along with it ultimately because we had settled here, and I realized that the job market was pretty bad. And I was pushing forty and I wasn't going to keep line cooking in restaurants. It seemed like a different type of opportunity. It sounded like, oh, we're doing something good where you grow all this food for the food bank and maybe it will be more viable. [Aaron] had wanted to sell everything—which he mostly did from the farm. I insisted on keeping the tractor because we'd paid for it, and move it here, and thank God we did. It was definitely a roller coaster. I moved here with some of my bakery equipment. I didn't have anywhere to store it. Ultimately, I got to the point where I needed to sell it. So, I sold all of it to another local bakery, and then all of a sudden we were building a kitchen and a bakery and a farm store—capitalizing everything again."

They began working on the food bank land with an unusual lease arrangement that called for Suzanne and Aaron to pay the annual rent in thirty thousand pounds of vegetables per season. It would go to the organization's Food Shelf and Meal Site program. They also set up their own farm store with a production kitchen and a bread bakery. They called the place the Kingsbury Farm Garden Market after the original owners of the land. Along with the veggies, they sold pickles, sauces, soups, tarts, full meals (from what they grew on the farm), and, eventually, bread. They planted berry bushes and cover crops. They set up three movable "high tunnels"—plastic-covered greenhouses to extend the growing season. There was also a three-bay garage building on the property. They decided to use one bay as the shed for the farm machinery, one for washing and processing produce, and one for a bakery. The food bank paid to pour a concrete floor slab and insulate it. They fixed up the old farmhouse for farm employees, but the rent went to the food bank.

It was a wash. They realized that the arrangement made it difficult to fire employees who didn't work out because they were living on the premises and you couldn't just evict them. To make things extra complicated, they faced a backlash from the local newspaper and their neighboring farmers who felt that the food bank was unfairly subsidizing the Kingsbury Farm operation, even though it was not true.

"The editor of a local paper was printing stuff that was just flat-out wrong," Suzanne said. "I remember there was actually a public forum we hosted at the farm and going off to the filing draw and showing them receipts from equipment we had bought just to prove that we had bought it."

Local people would come around and let their dogs loose to run around the farm, thinking that the Land Trust deal had made it a kind of public park. Persons unknown shut off the irrigation pumps from the river, though the farm was legally allowed to use them.

"Then things settled down. Lots of people loved the farm store," she said. "We ran it all spring, summer, and fall, and then in the winter we would open twice a week. When skiers came up we would put all this food away, like tomatoes and pesto and winter squash puree, and we'd make lasagna and pot pies that people could come in and get for the ski weekends. And that was just a brutal amount of work for not enough money. It was a one-person show. I was doing everything."

In late August 2011, Hurricane Irene ripped through New England—downgraded to a tropical storm when it got inland to Vermont. It did substantial damage to the Kingsbury Farm. Most of the fall crops were damaged and unsellable. The farm was effectively shut down for the rest of the season because the state was running excavators there to shore up the riverbank.

"That probably was the beginning of the end," Suzanne said. "Aaron dealt with that in a very different way than I did. And I don't think we ever really recovered from that. We ran the farm store for four years. Aaron's mom died, then his dad died. We ended up divorced and, while splitting up, had to finish a final season with the store open. After that, I didn't know what else to do so I kept baking. I figured I'd see if I could make a small wholesale business out of just baking bread and delivering. I ended up renting my kitchen space from my ex because he got the farm lease and all the equipment, and I basically got the house [up in Waitsfield], and I still needed to bake so I paid him to stay [i.e., to keep using the baking facility at the Kingsbury Farm] till I figured it out. I was pretty scared after getting divorced. I didn't see how I could possibly stay in a place like this by myself. I was pretty isolated. I baked, delivered, I did everything by myself at that point just to see if it would work."

It was a dark time for Suzanne. She was living alone in the renovated house up in Waitsfield, away from the Kingsbury Farm. But even renting the bakery space at Kingsbury Farm, Suzanne began to feel that she had a viable business going. She changed the name of the operation to Green Rabbit, originally the name of the farm back in New York State. She started thinking about building a new bakery next to her house in Waitsfield.

"There was no way I was going to continue in the garage at Kingsbury Farm. It was a very hard place to run a clean kitchen. There was a lot of dust from the farm. It was hard to control the atmosphere. It was hard to keep warm; it was hard to keep cool; it was hard to keep the humidity levels. I also knew that I probably wouldn't last five years of waking up at four thirty in the morning and driving to the bakery. The roads weren't plowed or sanded. At least twice I significantly went off the road on my way to the bakery in the morning."

It took six months of bureaucratic haggling with the town to get a permit to build a new commercial bakehouse on her property up in Waitsfield. She borrowed money from her parents to put up the building and again hired architect Danny Sagan to design it. On top of bread deliveries, there were gardens to build, a lot of heavy outdoor work planting soft fruit, looking after sheep, caring for forty pear, plum, and apple trees, and sundry other tasks. A new bakehouse must have seemed like a dream come true for Suzanne, but she continued to fret as the building took shape.

"It was kind of a push-me, pull-you dream because at that point I was over forty. I knew the reality of food business. I knew that it was just brutally hard and a ton of work. I often was feeling like I might be a little too old to bring it to this point—which I'm still wondering about. I have a business model that works so long as I am an integral physical part of it. The thing we struggle with is that we really can't afford to replace me unless we were to grow. And in order to grow, we need another van, we need another mixer—it's this damned-if-you-do, damned-if-you-don't situation. I've been looking into maybe changing the model. If we could do retail instead of wholesale—I mean, we'd probably always do some wholesale—but we'd have to be able to rely on more people. I'm the only one who feeds the starter. If something goes wrong, we have no bread. I can't put that responsibility on anybody else. And Clara [the assistant baker] is really good. I went through the first three years up here where every eight months my employee left. I would just get them trained and they'd leave. People think that it's really easy to replace a baker but it's not. It's the kind of baking we do, and the kind of skills that I need in someone, the amount of training I need to carry out to get them to the point where I can trust them. It's like hiring a finish carpenter versus just somebody who's going to frame and move on. Twice I had to go through Christmas and New Year's doing the whole load myself. And granted we weren't quite at this scale, but it still killed me. I mean I'd get to New Year's, and my elbow would be, like, out to here. It was just crazy."

To add to the difficulties, a set of brand-new bread ovens failed. To equip the new bakehouse, Suzanne had sold the old ovens at Kingsbury Farm and purchased a set of snazzy three-deck Gashor artisan bread ovens. They were made in Spain and cost $30,000.

"Nobody's making this stuff in America," she explained. "We started having problems. Our electrician noticed issues with the ovens even upon installation. The mechanisms to open and close the doors didn't function properly. We told the company, we called the distributor—they just kept laughing at us like we were making all this up. We were led to believe that if we needed service, there was service to be had in the US. Obviously, they were selling these ovens in the US, so somebody had to service them. But there was nobody to service them in the US. You had to go to Spain—people who didn't speak English—every time you had an issue. And they had actually gotten so bad we did have a little fire in one of the electrical boxes."

Suzanne was so vexed and so insistent that Gashor agreed to send an engineer from Spain to Vermont.

"They had no infrastructure here in the US, and they didn't tell anybody that. They sent this guy out who didn't speak English. I had to feed and drive him 'round for three days. I listened to him half knowing, half not knowing, what he was doing or what was going on. It didn't end up solving any of the problems. At this point I had started to get advice from my dad's friend, a lawyer in New York City. At first, he was just advising me through the process, and then it became a lot of advice, so we did pay him. But basically he said we'd never win a lawsuit. And I decided to try and pursue it through insurance. Well, then what happened finally—besides all the other problems we're having with it—in the middle of a bake one of the glass doors literally blew out like a bomb had exploded across the bakery. Normally we're standing right there. I just happened to have called away my baker to look at something on the other side of the bakery. I mean, it was borderline PTSD afterwards. We couldn't believe that neither of us had ended up being rushed to the emergency room with shards of five-hundred-degree glass in our eyes."

Suzanne said the manufacturer laughed off the incident. There was talk of arbitration—which had to take place down in the distributor's state of New Jersey—and which would require paying more lawyers' fees, so Suzanne declined and decided to pursue the insurance route instead. The insurance process took eight months and was like having another full-time job, with getting the documentation in order, filling out one form after another, consulting with a new electrician and engineers. Her insurer subcontracted out the claim to another insurance company. Meanwhile, she was down to running two, and sometimes only one oven deck. It went on and on. She was on the verge of giving up altogether.

"At that point I'm, like, if I'm going to keep the business I had to buy new ovens. Basically, my dad's lawyer friend became my voice of reason."

Suzanne borrowed once again and bought a new set of German-made ovens this time. They cost $40,000.

"I had said on the phone to the insurance company—and I was so sick of having these conversations over and over and over—I said, 'Listen, at this point, if you're going to send me a check for six hundred dollars, I'll just let you stop it right here because this has taken so much of my time, and I've already given in to the fact that I'm going to run this business till I sell it and then I'll make my money back.' And then she [the insurance rep] was like, 'Oh no, we're going to cover this.' And even then I was like, *Okay, I'll believe it when I get the letter.*"

The letter came. It turned out that the insurer covered not just the $30,000 loss of the Spanish ovens, but the additional $10,000 cost of the replacement German ovens.

"This was a crash course in life for me," Suzanne concluded.

When I visited Green Rabbit, the new ovens had been running nine months without any problems. Looking back on it all, Suzanne said the ordeal had felt "like a kind of martyrdom" in the cause of a career in bread making.

"It was kind of like I got on the train and I couldn't get off. And there were so many times when I was honestly so close to just saying *maybe I should just shutter it.*"

Now that she wasn't spending every spare minute on insurance minutia, she was focused once again on the business of running the business. I was impressed that in our extensive conversations, Suzanne never once fell into the precooked political rhetoric of the moment. She never brought up President Trump, or the race and gender issues that preoccupy so many people in midlife today—that are especially intense in the "woke" state of Vermont. She had been paying close attention to the particulars and problems of her trade above anything else for years. If she expressed any political viewpoint at all, it concerned the impediments and intrusions of government on citizens trying—against all kinds of tough odds—to run small businesses. And yet while this view might seem to be classically (capital *C*) Conservative, Suzanne did not place herself on that end of the political transect, and probably would have been surprised at the intimation. What bugged her were the payroll liabilities, the taxes she pays that subsidize big agri-biz, the extreme overregulation of American business, the many illogical prohibitions on what employees could or couldn't do, along with the insanely screwed-up medical insurance system that made it so hard for ordinary people to take care of their health—and her role in that as an employer.

In the spring of 2018, with her logistics finally operating smoothly in the lovely new bakehouse, she had more orders for bread than she could fill.

"We just started getting requests from all over the place," she said as we sat at her kitchen table in the renovated farmhouse, with the world flowering outside the bay window. "It's a rural state and we have to make the delivery routes work. If somebody wants you to go in this direction and they only want to buy a hundred twenty dollars' worth of product, you have to make it worth it, if you can't justify enough stops—again, all of this is so much to figure out. We've gotten much better at it. Like, we *did* go to Middlebury for a year and realized this is insane. We're driving so far and we're not making enough to justify that route. So, we stopped going to Middlebury and we replaced that by going south towards Woodstock [VT], which seems to be working. It's the volume that you can move. We can sell between four and seven hundred dollars' worth of bread. Just driving in that direction I'd say—I'm just trying to think—so come summer we'll be doing three-fifty with the Woodstock farmers' market, we'll be doing two-fifty with Clay, two-fifty with Hasso. We actually turn that into like six stops. So that would be a thousand dollars in one direction. And then I have Cam, my other employee, who drives my car in another direction. And so that's it."

That was it. Except for the final question: *Now what . . . ?*

Where does this country go from here? What are the forces in the world that determine the next chapter of history? Where does Suzanne fit in?

"I think people need to live far more humbly," she said. "That should have happened a long time ago, probably at least twenty years ago. Like, we have neighbors that work for efficiency advocacy groups, yet they fly all over the world for vacations. There's just this disconnect. Or I guess you call it greenwashing in your own mind. A lot of people genuinely don't even think about it and don't even know, so they're not even doing it hypocritically. They just honestly don't even know how much energy something like flying takes. I think the population has to decrease. I think more people need to have small gardens—and I don't know if this is possible. People join CSAs now [Community Supported Agriculture]. When CSAs started it really made sense. Like, the true philosophy behind CSA is that a certain region basically subsidizes a piece of land and a person who has the knowledge to grow food on it. And that piece of land provides that geographic area, or a certain population, with calories.

"Now," she said, "if that truly happened in its real sense, I think it would be a lot easier to localize something like food production and provide enough calories for people to live on. But it's really just a subscription program the way it functions. And I think if it actually was carried out in its true sense that we'd have less food insecurity in given regions as long as you still had a growing season, and fertile soil to grow on, and water."

"People need to be prepared for big storms," she resumed. "Everybody who can afford food should always have food and water and batteries. That's just a no-brainer at this point. Every time we see a big storm coming on the weather, I'm filling up pots. After [Tropical Storm] Irene, I take every threat seriously and I start getting butterflies. I had gone to see my mom and went to help out at a distribution center after [Hurricane] Sandy in the New York area. And it seemed an area like this [Vermont] is much more suited—I mean, it can cripple us, it can bring us to our knees and stop everything. But people have the tools and the knowledge to get through a day. And maybe you can't get to public school but you can start a fire. You can keep warm. You can cook on a grill. There's water to be had. You can make it potable. Your environment can be healthful instead of toxic. I mean, if you think about Long Beach [Long Island] or suburban sprawl being brought to their knees after Hurricane Sandy with no refrigeration and no electricity and no sewage, it's much more toxic.

"The thing about rural life is, yeah, we don't have public transportation so we're kind of forced to drive. I have this conversation with Clara, my young employee, all the time. She gets really frustrated when people talk about not owning your individual car. She's like, 'How could I work? You know, how could I possibly work? I have to work?' And yet you look at us, we're driving around all over. I mean we joke about just driving downtown with a horse, you know."

"Do you think this country can keep its shit together?" I asked.

"No, I don't. I don't think a federal government can serve the fifty states anymore. But I also don't think fifty separate governments will work either."

We'd covered the territory of Suzanne Slomin's long and winding road to a career that she cared deeply about, and the building of her business. She impressed me as remarkably stalwart and brave, a person who took full responsibility for what happened to her and who had navigated through considerable travail in the difficult task of shaping her own life her own way. I was grateful that she let me into it and trusted me to write about it. She graciously put me up for the night in the guest quarters in the new bakehouse, and the next morning when I left, she gave me many loaves of her excellent Green Rabbit bread. I ate most of one loaf before I got back over the mountain, heading west to the Champlain Valley.

Chapter 9

THE TRIALS AND HEARTACHES OF A GEN XER

I first encountered KMO a decade ago at an annual meetup of the Congress for the New Urbanism (the CNU). This one happened to be in Austin, Texas, at the Convention Center, a god-awful pharaonic-scaled monstrosity built to hold many thousands, where we few hundred New Urbanists rattled around like BBs in a packing crate for the weekend. It was a bad choice of venue—in a desolate part of the city where walking to anywhere else was like reenacting the Bataan Death March. But Austin was a "hot" city, economically and demographically, and the NUs were probably looking to hustle some work there, thinking strategically to counter the suburban sprawl development that was Standard Operating Procedure there.

For those who don't know about the CNU, the org was founded in 1993—the same year my *Geography of Nowhere* was published—by a group of then young architects, urban designers, developers, and even government officials who were dedicated to reforming land development practices in the USA away from sprawl and toward compact, walkable, mixed-use, human-scaled towns and neighborhoods. They'd had a lot of success since the 1990s putting up exemplary new town projects like Kentlands, Maryland; Seaside, Florida; and I'On, South Carolina; and also in reforming zoning laws in older established towns all over the country so that building a better alternative to the sprawl template became legal.[39] To me, they were one of the very few orgs in the land who were actually accomplishing something to prepare America for what might follow the collapse of the hopped-up cheap-oil economy.

So, KMO turned up in Austin and we made friends. Everybody in KMO's world calls him KMO—*Kaymo* is how I say it—but his proper name is Kevin Michael O'Connor. KMO, then around forty, was just setting up in a new podcasting career at the time. He was excited about it after two tumultuous decades of switching careers and many drastic geographical relocations. He had started a website called the C-Realm ("C stands for Consciousness-*ness-ness*," the echoey intro to the podcast put it), where he was entertaining scenarios that jibed with the possibility of economic collapse. He was following blogs like mine, and he'd come to Austin to record interviews with me and with several of the CNU's leading lights.

That very moment, April of 2008, events seemed to be shoving the world toward thresholds of criticality, especially the global banking system. A month earlier, the Wall Street firm Bear Stearns collapsed, presaging the worst financial crash since the Great Depression in the fall of that year. The housing market was cracking up catastrophically, along with the many bundles of "innovative" bonds called mortgage-backed securities—composed of mortgages held by Las Vegas busboys and Phoenix hair stylists, who suddenly couldn't keep up the payments on million-dollar tract houses they'd hoped to flip for a fast killing. The Frankenstein derivatives of these securities included the phony-baloney credit default swaps that had been concocted as "insurance" against those janky bonds. These would go south too. Plenty of ordinary folk who were not engaged in house flipping lost their homes. Also that spring, the price of oil was zooming ominously toward a summer high of $140 a barrel, an economy killer, since just about every activity in modern life required cheap oil and suddenly it wasn't cheap anymore—and in the coming winter, the price would crash so hard that the oil companies couldn't

---

39 Zoning laws had evolved since World War II, and then ossified, to the point that suburban sprawl was the virtually mandated outcome of land development all around the USA.

cover their costs of production. In the spring of 2008, you could see trouble coming from miles away.

Though KMO was new at podcasting, I was impressed with his professionalism, his preparation, and the excellent questions he asked when I submitted to a sit-down session with him in Austin. I stayed in touch with him after that, became a fan of his podcast, and was an interview guest at regular intervals in the years ahead. As a science-fiction fan from boyhood, and then as he came of age in the early 1990s, KMO was an early adapter to computer technology and all the doors it promised to open. But he became increasingly disenchanted with its techno-utopian promises, especially as the crisis I'd named *The Long Emergency* rolled out after the 2008 crash. KMO himself had lived through a series of personal vicissitudes that made him a kind of poster boy for the social and economic changes that Generation X faced—all summed up in the emerging realization that they would not share a ride on the American Dream bus that the Boomers had coasted along on for most of their lives.

KMO was a true inquiring mind. As a Gen Xer, he still shared some of the old lingering hippie spirit, along with a balancing touch of gritty, nomadic adventurism. He'd traveled more than once to the Amazon jungle to take part in Ayahuasca sessions, a brew of the chaliponga vine (*Diplopterys cabrerana*) and other herbals with potent psychedelic alkaloids related to the mind-bending drug DMT (N,N-Dimethyltryptamine). His sci-fi pursuits had led him to an interest in dystopian futurism, especially zombie tales and movies, which seemed to him an especially vivid metaphor for the slow-moving train-wreck effects of industrial society he saw happening all around him. But he also trafficked with other emerging phenomena on the podcast: permaculture, peak oil, the Singularity, space travel, climate change, the Matrix, and the overarching need to chart a new course through a scary and mystifying future.

When I met up with KMO at the Austin CNU, he had been living in his native Arkansas for a while, as we'll see. But soon he was out of there and turned up living down in the Eastern Shore of Maryland. He'd gone through another round of extraordinary changes, including a split from his wife, with whom he'd had two children. KMO was working a call center job there to pay child support while struggling to keep podcasting; fans were beginning to make dollar contributions to that endeavor. In the decade after that, he moved again, this time to a fabled old hippie commune in the backwoods of Tennessee, then went on the road—an extended couch-surfing speaking tour for a book of C-Realm podcast interviews titled *Conversations on Collapse.* Then he sojourned for a couple of years in Brooklyn with a girl he'd met among his podcast fans, and finally went up to a small town in Vermont on the Connecticut River. He was only two hours from where I lived, next door in easternmost upstate New York, and I started seeing a bit more of him, including the day we sat down to talk about his life for this book.

BELLOWS FALLS

Bellows Falls, Vermont. It sounds like the place where Jimmy Stewart's character lived in the classic 1946 movie *It's a Wonderful Life*, and maybe it used to be something like that—but, really, so did just about every Main Street town in New England before American industry got on a slow boat to China. The Connecticut River, which is the boundary between Vermont and New Hampshire, runs through a severe rocky flume there, and the raging water powered an array of paper mills, while the first canal built in the USA was dug to bypass the falls for boat traffic. It took from 1791 to 1806 to complete the task. Later, railroads ran up and down the Connecticut Valley.

Spring had not quite sprung when I drove over to Bellows Falls in early April, and the tedium of endless winter formed a grim patina over the place that underscored the decades of disinvestment in its townscape. The business district was an ensemble of handsome red brick buildings, including an Italianate opera house and town hall with a campanile, a Greco-Roman Carnegie library, a Spanish Colonial Revival post office, and the Windham Hotel, which for years housed one of Vermont's only gay bars. A lot of good old buildings had been lost over the twentieth century and replaced by the ghastly products of the commercial construction industry: one-story, tilt-up, stuccoed sheds with crappy windows and strange rehabs favored by the no-skill-necessary Do-It-Yourselfers. Many grand Victorian homes had been built by the town's nineteenth-century men of business, and they still remained standing, though not a few had turned seedy one way or another, cut up into apartments or converted to funeral parlors. KMO had been living in a little cottage just where a town street turned into a country road. The house was owned by his Brooklyn girlfriend, but they'd separated after several years and KMO had moved to a garden apartment a few blocks away with his old three-legged cat, Mocha.

He was a few weeks shy of his fiftieth birthday and he looked great—lean and ripped. He'd been getting a little chubby over the years, but that was gone now. He was working out a lot in the gym and doing yoga. Lately he'd gotten on a ketogenic diet (a.k.a. paleo diet), eating few carbs and a lot of meat. It had firmed him right up but it was expensive to eat that way, and KMO was one of those struggling Americans who could not afford health insurance. Eating right and staying strong was his health insurance, he said. The apartment, in a boxy modern building of about ten units, was comfortable, but it showed all the signs of a grown man living alone. Stuff was arranged mostly for sheer utility, and there wasn't much in the way of decor. We settled in at his kitchen table.

KMO was born in Fayetteville, Arkansas, where his mother's family came from. But he spent most of his childhood in Kansas City, Missouri. He had one sibling, a younger brother, whom he described as "one of those unfortunate 'failure to launch' cases." His father, originally from New York City, worked as a Secret Service agent.

"The Secret Service is a branch of the Treasury Department, so they hunt down counterfeiters and check forgers and credit card fraudsters," he told me, explaining why his dad was not posted in Washington. "You had to accept the transfer in order to get promoted. They just shuffled people around hither and thither. They also protect the president, the vice president, their families, and so not all these guys actually are around the president all the time, but he did all that stuff. He spent time on the campaign trail with Ronald Reagan. Actually, he was present when Ronald Reagan got shot. He spent time with Nixon and Ford and with one of Ford's sons, and for several years he was stationed in Independence, Missouri, guarding Bess Truman. That was an easy gig. She didn't get out much."

The family lived in the suburbs, and KMO was happy there. For all its peculiarities and shortcomings, let's not forget that suburbia was so hugely successful because by and large Americans liked it, and within the context of the twentieth-century US economy, it worked. In the 1970s and '80s, when KMO was a kid, that economy was just entering its long, painful decline.

"It was pretty great," he said. "I spent my days on my bicycle out with my friends, getting into misadventures and had many a brush with death. I think about it now and I realize how lucky I am to be alive. I got hit by a car one night when I was on foot. I was with my father. I had just seen a movie, and the parking lot was jam-packed as we were trying to leave. The movie theater was just a couple of miles from my house. I said, 'I can get to the house before you can.' And he took my bet. Within thirty seconds I had to cross a major highway and looked both ways and somebody saw that I was waiting to cross and he said, 'Go ahead kid.' So, I ran and I was almost all the way across the highway and I looked to my right and headlights came on right next to me. I sort of flinched and jumped a little bit, and the car hit me in the hip, and I went up over the top, and I broke my arm when I hit the ground but didn't realize it at the time. I was in total shock. Some other people saw it happen and prevented me from running off, which was what my instinct told me: to just run home. They took me back to my dad who was still stuck in traffic trying to get out of the movie theater parking lot, and he drove me to the emergency room. But had I not jumped up a little bit when the car hit me—it was going forty-five miles an hour—I would've been killed."

His parents separated when he was thirteen and divorced four years later.

His father turned down promotions in the Secret Service because he would have had to accept a transfer to another city, and he wanted to stay near his boys. He also made an enemy of his boss and his career stalled. He retired as soon as he hit the twenty-year mark and took a job as head of security at a hospital. Later, he worked in code enforcement for the city of Lee's Summit, Missouri, a Kansas City exurb.

KMO did not thrive academically, calling himself "a bad student, indifferent." He followed his own interests. He liked drawing cartoons, making comic strips, which he would return to as a serious venture later in life. He was in the high school art club and took a lot of art classes.

"I liked to read science fiction, so I took classes where I had to read books and write about them and demonstrate my knowledge of what I'd read. Those, I did well in. Math classes, science classes, boring classes, I just didn't really bother. I think I had a C-plus average. I actually hated high school."

He didn't want to go to college, just wasn't interested. Instead, he answered the come-on of an enlistment officer to join the Marine Corps. When he went to sign up, they shunted him into a delayed-entry program.

"It's now the summer after my senior year of high school. I have no plans to go to college. And I'm just waiting to ship off to the Marine Corps. But I have a summer job. And the Marines tell me, 'Okay, you're shipping out on Friday.' So, I quit my job and I gave stuff away. I was going into six years active duty. I was gone. This was 1986 and I got down to the Military Entrance Processing Station, the MEPS center. I went through their fourteen-hour ordeal of physicals and medical stuff. And I'm sitting on a bench waiting for a bus that's going to take me to the airport, which is going to put me on a plane and send me to San Diego, and somebody comes up to me and says, 'Are you O'Connor?' I said, 'Yes sir.' He says, 'Your program is not on the computer today,' which means if I were to go to boot camp, I wouldn't graduate at the right time to go on to my chosen occupational specialty school. So, the next day I'm back at my mom's house. I don't have a job anymore. I'm not really doing anything. Then I get mono and my recruiter tells me, 'Don't tell them you got mono.' They pick a new date for me to go, and I go down to the MEPS center again. I do that whole ordeal again. They're asking me: 'Do you have this, this, this, this, or this,' and one of the things they asked me about was mononucleosis. I kind of hesitated. They picked up on my hesitation and took me off into a side room. A doctor and an officer said, 'If you've had some illness and you don't tell us about it, you can go to Leavenworth prison for a fraudulent enlistment.' Okay, I had mono and they said, 'When?' and I said, 'I think I still have it now.' And they said, 'Okay, you're disqualified for six months.' Now I'm sitting around doing nothing with six months to kill. So, I enrolled in a community college, and I took nothing but art classes and art history classes and introduction to philosophy. I didn't think I was going to be there to the end of the term. I didn't care about grades at all. I was just doing what I wanted to do. I got all As, and toward the end of that first semester

I decided, *Oh, this is not bad. I think I'll do this.* By then my [Marine Corps] delayed-entry contract expired. They wanted me to sign a new one. I said, 'No, I don't think I'm going to.' I was kind of off the hook but they didn't make it easy. They were real jerks about it. But yeah, essentially I didn't have to go."

He knocked around a couple of community colleges, got As in his classes, and earned his associate's degree, then transferred to the University of Missouri, where he latched onto an opportunity to study in Japan for a year. Eventually he earned a degree in "general studies" with a triple major in philosophy, fine art, and East Asian studies. Then he was off on a series of adventures. After graduating, he spent the summer with the Alaskan fishing fleet.

"I was on two boats there. The first was a fairly large processing ship that was at anchor in Bristol Bay. The smaller boats would actually catch the salmon, would bring them to us. We would take them on board, process, freeze them, and then transfer them over to a Japanese freighter, which was anchored right next to us. That's pretty hard work, fishing is. Then the ship went to Dutch Harbor, Alaska, which is just a fishing port with an airport halfway down the Aleutian chain. Some guy walked into the room where I was sleeping, and he was watching this other guy pack up his stuff, and I woke up and asked what was going on. This guy packing up was about to jump ship and go work on this cod longlining boat. And the guy I didn't know was the one who had recruited him. The guy recruited me to go, too. So, with no warning or time to think about it at all, I made a very impulsive decision. I went from the processing boat to a small longlining operation. Well, it's not only hard work, it's dangerous work. And it was miserable work. I got paid by the hour on the processing ship. On the longliner I was supposed to get a percentage of the catch. And I got ripped off. I got nothing on that second job. I worked my ass off under dangerous conditions that summer. Right after the fishing fiasco is when I went to Japan for the second time."

He taught English in the city of Nagoya, made a nice chunk of change doing it, had a lot of fun in a different culture, and managed to pay off his credit card debt.

"Then I decided to come back and go to grad school for philosophy," he said. "Back to the University of Missouri at Columbia. I spent two and a half years there. I had a thesis that I had written but I never assembled a thesis committee. I never defended it. And I had a couple of courses that were incomplete that I never picked up. So, I did the work for a master's in philosophy, but I don't have the degree. The only reason I could stay there and do that was I got a tuition waiver and a stipend because I was teaching undergrads. There was a time limit on that and I had exhausted it, basically. So, I had to move on in order to make some money, and I had some friends who are moving to Seattle. I knew a guy who had worked at Microsoft early on and made a lot of money. I thought if I went out there, he'd get me a job at Microsoft, and I went out there and he said, 'Oh you should go talk to Jeff Bezos over at Amazon.com.' I had three interviews at Amazon. The final one was with Jeff Bezos. That was 1996."

At the time, Amazon occupied one floor of an office building in downtown Seattle with one warehouse on the south side of town. KMO was hired to do customer service.

"Marketing was in the room next door to a conference room where Jeff was always talking with investors and the other executives at the company, and he has this big, booming laugh, and he was always laughing. We'd be on the phone or just answering emails from customers and Bezos was right next door presenting his vision to the people with the money and just laughing, laughing, laughing."

KMO described him as "a very charismatic person."

"I thought I was going to make a lot of money," he said. "I thought the stock value would increase. I had good options. I wish that I had just exercised them and held them. But after two years there—in the fall of 1998—my father's second wife left him and he committed suicide. I was still working at Amazon, and I was pretty devastated by that. So, I quit."

KMO had clashed a lot with his father when he was a teenager—what teenage boy doesn't?—but by his late twenties, he was past that. It also happened that in two years with Amazon the value of his stock options had zoomed. After quitting, he called his stockbroker and sold. A few days later, $650,000 landed in his bank account. Within a couple of months after he left the company and sold his stock, the price of Amazon stock tripled.

"If I had held on to it, we probably wouldn't be talking right now because I would've taken a very different course through life. I remember the day that I sat out at a table in front of the deli across from where I lived and I wrote one check and paid off my student loans. A good day. But this is also just after my father had committed suicide. So, those initial years when I had all that money from Amazon were some of the worst times of my life. I was just really torn up about the death of my dad. I did see a therapist."

I suggested we go get some lunch. It was chilly and nasty out, and there was still some of the frozen slush and snowplow debris that Vermonters call "snard" along the streets so, rather than walk, we drove to a coffee shop downtown. There were few people on the street there, an air of abandonment and desolation, as though the town was only 23 percent alive. The coffee shop seemed to be struggling, too, with a sparse lunch crowd, more evidence of the persistent economic malaise in Flyover America. KMO was subdued, likely from revisiting painful memories of his father's death. We didn't talk much while scarfing down our sandwiches. But when we returned to the recorder after lunch, KMO perked up and became voluble again.

A GEN XER OUT IN THE REAL WORLD

"The first year that I was away from Amazon," he said, "I spent about six hours a day drawing comics. I had a really clear vision that I wanted to be a cartoonist. At the same time, I had money to blow. And Seattle's a nice place to live and have money. I was partying and, you know, lots of good sex."

His girlfriend became pregnant.

"We had an appointment to go get an abortion. And I was the one who said, 'I don't want to do that. I think we should have this kid.' So, we ended up getting married and had another kid. But then the money was running out. I had bought a condo in Seattle and sold it. We moved out to the Olympic Peninsula and lived there for a couple of years. And then as the money was running low, we sold that house and moved to Australia."

They landed in Perth. The city of Perth, at the farthest southwestern edge of Australia, is reputed to be the major world city farthest away from any other civilized place on Earth.

"I kept working on comics, went to the gym a lot, drank a whole lot," he recalled. "We looked at starting a business. There was an organic fruit and veg store well south of Perth, a couple of hours south, but in a lovely place, a lovely little town. It was for sale and I had the money to buy it. We were thinking about that for a good long time and some friends of ours in that area said, 'We have an accountant in Perth. You should really take this business opportunity to him and have him look into it.' And I'm so glad we did because the accountant uncovered the fact that the profit the business was showing was a fiction. They were sliding into debt to their suppliers. It would not have worked out."

Their stay in Australia lasted only six months.

"I decided at that point that I wanted to move back to Arkansas and buy some land and start a farm," he continued. "My family's in Arkansas, my mother's side. My father was an only child. He was from Brooklyn, the East Coast. Both his parents were dead by that point. The place where I moved to, Berryville, Arkansas, near Fayetteville, is very rural, although it's ten miles away from a town called Eureka Springs, which is, if you're gay in Arkansas and you can't escape to the East or the West Coast, you go to Eureka Springs. There is a writers' colony there and there's an annual digital film festival. It's a tourist destination with a Christian theme park, so there's a great passion play there. They have a weird combo. There's also an annual motorcycle gathering. In the summer you get the roar of a thousand engines, and there are roadside lodges that cater to motorcyclists in the same town where you've got the Christian theme parks and the passion play and the museum that demonstrates that there were dinosaurs on Noah's Ark."

After the family landed in Arkansas, KMO met a Frenchman there, a former commodities trader named Patrice, who had chucked it all to become an organic gardener. He had three-quarters of an acre on the outskirts of Eureka Springs where he had built a market, and he appeared to be supporting his family with it. He was also selling to some restaurants and at the local farmer's market. KMO made a deal to apprentice with him and studied under him for the better part of a year in his quest to become an organic gardener. But by the time he was prepared to start building the necessary farming infrastructure on his own property, the last of his money stash was gone.

"I bit off more than I could chew in terms of building a house with a kind of a death race against the bank," he said. "I had to get a job, and the jobs to be had in Berryville are few, and the returns are meager. So, we moved over to Fayetteville and I started selling insurance. I met unsavory characters and I saw unsavory practices and got exploited. But then I just happened to be working in the senior market in 2005 when Medicare Part D rolled out. I had manned a booth in a Walmart and all the seniors brought me their list of medications, and I would go through the books and go through a computer program and figure out which Medicare Part D package was going to be best for them, and I made a bunch of money in a short time doing that, which was kind of an unfortunate thing. I got that money and paid off a lot of debt with it, and I decided to buy a house on a farm. This is late 2006 and this is when the mortgage racket, which would blow up two years later, was in full swing. I did not qualify for a mortgage in any sane way. But at this moment in time lots of people wanted to give me a mortgage. So, we bought this five-acre farm and a nice house and I tried to make a go of it."

He described his organic gardening endeavor there as "minimal." He kept some chickens and grew some vegetables, trying to emulate a few things he'd learned watching Patrice back in Eureka Springs. In the meantime, he had discovered the new communication form called podcasting and was spending more and more of his time at it.

"I knew I was never going to make a living on that farm," he said. "We had an outrageous mortgage, and basically blew through the windfall from the insurance gig, and lost that house and property to foreclosure in 2007. I was still doing insurance work. That was where my income was coming from. Then we moved to a rental house nearby. I really got into podcasting, and the audience growth was really fast. I projected that off into the future and I thought, *Oh, wow, this is going somewhere!* I really stopped trying to make a living in insurance so I was focusing on making a living podcasting and my wife did not believe it was possible. That's when the marriage broke up. We continued living in one house, but in her mind, she was already divorced. Then she convinced me to pick up and move to Maryland because she had family there. When I got to Maryland, that's when I got my job at Comcast doing customer service again. One day I was at work and I got a text message from her saying that she had taken the kids to a location she would not disclose and that the marriage was over."

They'd barely just moved into a big house there: KMO, the two kids, his wife, her brother, and her mom. He got the *Dear John* text at the Comcast call center, went back to the house, and discovered that most of the things that had any value were gone, along with his wife and kids. KMO was aware that they had not been getting along for quite a while, but it was still a shock to know it was really over, especially when he learned that she'd been conducting a romance on the side with another man—whom she then moved in with. For about a month, KMO lingered in the big house with his wife's mom and brother.

"I looked around and I found a place where I could go. It was the Elk River House, this two-hundred-year-old former hotel right on the Elk River, which is at the top of Chesapeake Bay. The other people living there were adults whose lives weren't going the way they expected. But then there were a lot of rich people living right around them. The house next door was the weekend playhouse owned by the business manager for Madonna and U2. We learned that the neighbors called the place where we lived 'The Halfway House.'" At that point, my life really seemed to track the fortunes of the country. I got rich during the dot-com boom. I lost a house during the housing crisis. And I'm struggling to make ends meet in the so-called recovery, which, you know, was a recovery for some but not for most."

KMO was not getting rich podcasting, but it did bring him into contact with a lot of interesting people. One of these was Albert Bates. Born in 1947, Bates was an early member of the archetypal hippie commune known as The Farm in Summertown, Tennessee, about seventy-five miles south of Nashville. The Farm was organized in 1971 under the charismatic leadership of counterculture activist and guru Stephen Gaskin (who died in 2014), a self-described "professional hippie," who led three hundred followers from California to the almost two-thousand-acre settlement in rural Tennessee to found an intentional community based on pacifism, veganism, marijuana, and the renunciation of personal possessions. Gaskin's wife, Ina May

Gaskin, would become a pioneer of modern midwifery and publish many books on the subject. The Farm was the model for scores of similar hippie establishments around the country. The communal arrangements there lasted until 1983 when an epochal shift called "The Changeover" happened, the equivalent of the fall of communism on the small scale. Henceforward, Farm-dwellers had to earn their own income, support themselves, and pay dues to live there, while still carrying on many of their eco-progressive political endeavors. Quite a few of the hard-core hippies just left. The administration of The Farm's property upkeep and business interests shifted from Stephen Gaskin to Albert Bates, a former attorney who had argued environmental and civil rights cases in the Supreme Court. Bates rebranded the enterprise as the Global Village Institute for Appropriate Technology—the Ecovillage, for short—as a sort of small-business incubator. Decades later, in the course of things, KMO happened to interview Albert Bates for his C-Realm podcast, and a bond formed between the hippie elder statesman and the Gen Xer that would open up another door for KMO—a door into a one-room cabin in the woods.

"After my wife took the kids and moved out and I was living in the Elk River House, I was working tech support for Comcast in Newark, Delaware, and I hated that job. Eventually, I was fed up. I was, like, forget it . . . and I quit. That was in December of 2009. I had assembled this couch-surfing speaking tour with somebody who had been a guest on the program many times named Neil Kramer.[40] We did a few gigs here on the East Coast and then we went up the West Coast. We started in San Diego and we ended on Orcas Island in Washington State. I really had no idea what I was going to do at the end of that trip, and very near the end, I got an email from Albert Bates inviting me to come to the Ecovillage Training Center and be the winter caretaker there.

"By the time I got there, it was March in Tennessee. Winter was basically over and [Albert] looked around for different roles for me. In the end

---

40 According to his website bio (http://neilkramer.com/about/), "Neil Kramer is a philosopher and teacher. His work focuses on spirituality, mysticism, and metaphysics . . . [working] with people to advance development through a deeper understanding of self, soul, equilibrium, divinity, and transformation."

I was basically just podcaster-in-residence. He gave me a place to live and paid me a monthly stipend, and it was enough for me to keep paying my child support. I spent two years living in a cabin in the woods in Tennessee, with all of these young, idealistic hippies coming through, learning organic gardening and alternative building techniques."

Ironically, while he was podcasting about the revolution in high tech, peak oil, and other cutting-edge issues, the Wi-Fi at the farm was so inadequate that he had to drive twenty miles round trip to the nearby towns of Hohenwald or Lawrenceburg to get on broadband powerful enough to post the podcasts to his website.

Stephen Gaskin was still around, near the end of his life, a kind of revenant or phantom presence haunting the place.

"When a man is in a position of social prominence and then loses it, it takes a physical toll," KMO observed of The Farm's deposed founder, "and when I was there, he was in terrible shape. He was a very tall guy, and he had been physically imposing when he was younger, but he was just a walking skeleton."

While hanging out there, KMO collected his recorded podcast interviews into a book he published called *Conversations on Collapse*, which prompted him to contemplate launching another couch-surfing tour to promote it. During that time, he was contacted via email by a podcast fan in Brooklyn named Olga Kuchukov, a massage therapist. He asked her to look into finding a New York speaking venue for him. Soon, he went up there and they met in person, and a romance started.

"She actually was going to sell her place in Brooklyn and move out to Tennessee," KMO said, "and we were going to build a place in the woods near the farm. Fortunately, we didn't do that. It would have been a fiasco. That old temptation came back to own a piece of land. Instead I moved to Brooklyn."

KMO went from a one-room cabin in the woods with no plumbing to Olga's fifth-floor Brooklyn walk-up, with Mocha the three-legged cat he'd adopted on The Farm. It turned out he loved Brooklyn, and he was in love, and he loved being in love. He was working in a yoga studio, making some money, putting up podcasts (which also produced some income), drawing cartoons, and enjoying the multitudinous cultural offerings of the big city after years spent in lonely American flyover backwaters. He and Olga lived together in Brooklyn for five years.

They began planning for the future. Of course, KMO had developed a rather dark view of the future, a combination of his personal experiences with all kinds of loss over two decades—money, family, property, prospects—and he stayed tuned in to the economic collapse network on the web, plus what he was observing on his own about the growing economic, cultural, and political dysfunction across the nation. I asked him to summarize it.

"I guess it would be that the prosperity the United States enjoyed through the twentieth century was largely a product of abundant petroleum, and being the last man standing after World War II, and basically dictating the terms upon which the rest of the world got to rebuild that was granted through the '40s, '50s, and '60s. As I look back on it now, I think it really started to unravel during the 1970s and continued to unravel through the '80s and the '90s with distractions like the bubble around tech stocks, and then the housing bubble. I had this narrative of a succession of bubbles. Each one appeared to be the beginning of a new era of prosperity. But each one was a scam, essentially a Ponzi scheme that left a few people richer and the vast majority of people ultimately poorer. And that with each recovery, fewer people would recover. And that narrative seems to be holding."

I asked him: "Have you continued to be worried about the trajectory that we are on?"

"I'm not worried," he replied. "I say I'm pessimistic, okay, maybe fatalistic. I think it was in 2008 when I first talked to John Michael Greer and understood his story of catabolic collapse where we have a partial collapse, and then a partial recovery, and a partial collapse, and a partial recovery.[41] That makes sense to me. And I think that there will continue to be periods that look kind of bright and which will be spun as the beginning of the new age of prosperity for the United States. But none of them will work out."

---

41 John Michael Greer, blogger at *The Archdruid Report* (discontinued 2017) and now at the site www.Ecosophia.net, author of *The Long Descent* and many other books.

Being a fan of KMO's podcast, and listening to the various collapse-niks over the years, Olga apparently shared his view that the future looked precarious. She owned the apartment in Brooklyn, where the property market had stayed red-hot, and was worried that yet another real estate bubble would pop and deflate its value. She wanted to sell, lock in her gains, and move out of the city. KMO still liked city life, but he also had Vermont on his mind.

"When I was on that couch-surfing book tour back in 2010, [friends] in Bellows Falls got me a speaking gig at the library down in Putney, which is fifteen minutes south of here. And more people showed up to that than showed up to my talks in big cities. I thought, *Wow, if I can pull an audience like that here then this is a place that the people are tuned into the things that are important to me.* It just seemed like a good cultural fit. We made many visits here before we seriously considered moving here. So, we had contacts here and we had a vision of what it would be like to live here."

He also had the opportunity to do a radio show on the local station in Bellows Falls.

"The radio station, where I still have a weekly radio show, was a big draw for me. The fact that I could take all that podcasting expertise and just apply it to terrestrial radio, it's something I had wanted to do for a while."

Olga went ahead and purchased the cottage in Bellows Falls. She'd had a notion about living way out in the country, but KMO persuaded her it would be better to be able to walk easily into town. They made the move in 2017. It was problematic from the get-go. The relationship was fraying the last year they lived together in Brooklyn. The move didn't improve their relations. Olga was driving down to the city frequently to make money doing massages for clients down there. She owned the cottage, and KMO could not make decisions about it. I drove over to see him in Bellows Falls and be a guest on his radio show around that time, and it was obvious that they were at odds. Not long after that, KMO had to find a place of his own, which turned out to be the apartment in which we were recording his story.

He was about to turn fifty. His sons were seventeen and thirteen, and he took pains to see them throughout each year (as well as paying child support). He was keeping busy in Bellows Falls. That year, he'd acted in and helped write a dramatic video serial, *Strange Events at the Vilas Bridge*, made in town by the local TV station.[42] He was still doing his radio show, though not with the same gusto he originally felt. He was shooting video for a local TV station and working a few days at the gym where he pumped iron. He was still recording C-Realm interview podcasts, though he had jiggered the content of his website behind a partial paywall. He had returned to his love for drawing comic strips and was posting a serialized sci-fi fantasy strip that was also producing some revenue.

"I spent years and years taking art classes and actually doing comics and building that skill," he said. "Then I went for years not using it at all, and that was kind of eating away at me. So, I'm really happy to be doing it now. And I basically just want to sit here and get some work done."

I came down to the same final question I'd asked all the other people who had let me into their lives, and pretty deeply so, over the spring and summer when I was on the road: *Now what . . . ?*

"This gets into my hobbyhorses where I'm on the mike and talk about this stuff for at least an hour," KMO said. "I think that social media, and Facebook in particular, is driving us insane. I think it has been deliberately designed to hijack our neurological reward systems and keep us engaged in a way which causes people to be really vociferously ideological. There's no middle ground these days. If you're not in agreement with somebody then you're the enemy. I did vote for Trump. I think he's absurd. I think he's bad for the country. But I think that the mainstream Democrats have gone insane.

"I think we're certainly going to have a leftward swing. I think Trump has a good chance of getting reelected if he stays out of prison. But then I think in the next electoral cycle we'll certainly get a Democrat, and possibly even a Bernie Sanders–type character, somebody with that same agenda. We understand the concept of the erosion of the middle class. But the part of the middle class which has failed to, thus far, fall off the cliff, has drawn ever closer to the people with the money who they serve. They think all of this rabble-rousing talk about working people being left behind in the middle of the country, being neglected, that that's all just a cover for racism, and we

---

42 Part 1: https://www.youtube.com/watch?v=2z3zJsWvKPg;
    Part 2: https://www.youtube.com/watch?v=5yZ9BhPlyUI

hate women. I know that it's not appealing to most young people, the whole social-justice warrior thing aside. That's a fairly small contingent which gets more than its fair share of notice. I know my sons have no interest in that stuff and in fact I'm fearful that my oldest son is gravitating to what's known as the alt-right. He's a really smart kid, and he's smart in ways that I'm not. Like he's really good at math and science and computer programming, and he has a bright future, if current trends hold. But with the rapid advance of artificial intelligence, I think a lot of people now, a lot of young kids who think that they have a bright future in the tech industry, don't."

"Are you surprised at the delusional quality of our national life?" I asked.

"I understand," he said, "if you're doing well in a system which is collapsing, it's in the interest of your own psychological comfort to imagine that the collapse is a fairy tale, that it's really just lazy people who should get to work and stop complaining. I think our civilization climbed a lot faster than any previous one and it can go down a lot faster. I've been living through the collapse most of my life. And I think my kids will live through it for most of their lives."

And that was where we left it.

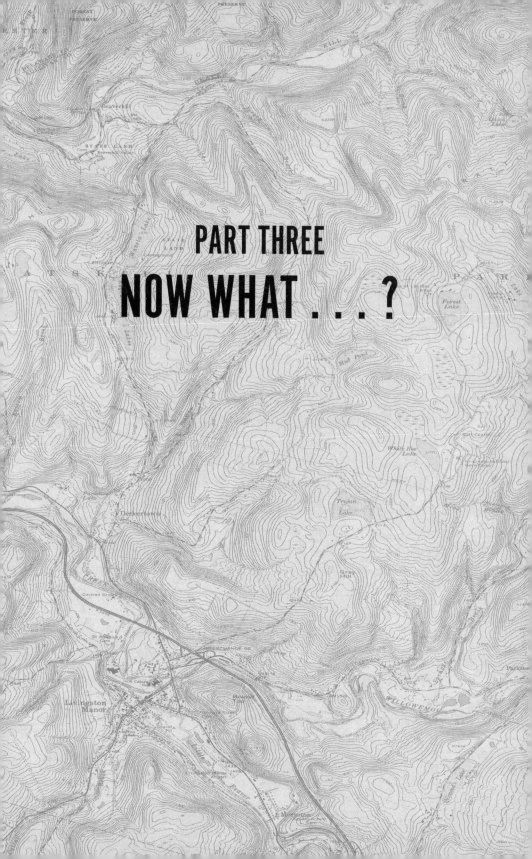

PART THREE
NOW WHAT . . . ?

The portraits of the men and women I've presented at the center of this book share many key qualities in positioning their lives for the discontinuities of society and economy that comprise The Long Emergency. One thing they have in common is a basic recognition that something big is up, that the path forward in human history will not be the business-as-usual *march of eternal progress* we've been accustomed to in recent generations. Sure, a lot of *ordinary* Americans feel anxious, angry, and depressed by what they see around them and experience directly themselves—lost incomes and vocations, pervasive dishonesty and racketeering, political idiocy, crumbling infrastructure, manufactured sexual strife, engineered loneliness, and lots more—but I daresay the general theme *out there* is a belief that the problems with our current arrangements can be "fixed" if we just elect the right leaders, apply the right "policy tools," guarantee equal outcomes, have faith in technological innovation, recycle more trash, and "celebrate diversity." They are hoping for the best, while living lives that make them frantic just getting through each day, and losing ground even so. The much more likely scenario ahead is that we will have to make very different arrangements for everyday life in a disorderly, impromptu, and emergent process, and that many of the comforts, conveniences, and certainties of past times will no longer be there for us.

The people I wrote about have been thoughtful and deliberate about the alternative lives they have chosen and have taken specific actions for practical reasons. One thing they surely agree about is that the scale of just about every activity or mode of organization running today is too large and will have to be reduced. Another shared trait is that they are *adapters*, not *mitigators*. This is a major theme of Part Three: that it makes more sense to roll with and adapt to new circumstances that reality presents rather than attempting to mitigate change with an overlay of additional new complexity on top of old, failing complexity in a quixotic effort to prop up the status quo—for instance, the wish to "fix" the problem of suburban sprawl by replacing individual internal combustion cars with circulating, driverless, electric cars. That scheme comes out of the folder I label "Techno-narcissism." It seems superficially appealing, yet the suburban-sprawl mode of living is a failed experiment for many other reasons that have nothing to do with what energy source or engine type makes a vehicle go or who is at the wheel.

The people I wrote about in Part Two are adapters, specifically *early* adapters. They share some cardinal virtues: self-reliance, grit, a clear sense of purpose, ingenuity, an array of skills, resilience, and, most of all, an allergy to conventional thinking. They're intellectuals in the broad sense of being engaged with ideas, but they are primarily concerned with getting things done in their lives, with taking action. These are characteristics that have enabled human beings to survive and thrive through all the ups and downs of civilization. In the late techno-industrial age, though, it's been possible for people to thrive by doing as little as just showing up—for an unsatisfying job, or an appearance at the social services office, or picking up a dividend check in the mailbox. It required very little initiative to stay in the game. The world population increased from one billion souls around the year 1800 to nearly eight billion today. (In my lifetime, the USA population more than doubled from 160 million to 330 million.) Those trends are sure to reverse, though even in periods of hardship people have sex and produce children, so the population may overshoot for a while even when the converging difficulties of The Long Emergency get traction and start to bite hard.

The implications are pretty unappetizing: namely, that when the population does start going down, it will be fast, hard, and ugly. The usual suspects will come onstage—starvation, disease, and war—and will do their thing, as we used to say. I don't want to dwell on that for reasons I'll explain presently, except to emphasize that it will take some initiative, grit, skill, and probably some luck to stay in the game, which is why it pays to be an early adapter.

Conditions will not be exactly the same everywhere on the planet. Some places will fare better than others, even if the overall trend is markedly down.

As a general rule one might assume that the damage to current living standards would appear at the margins first and work toward the core. But today, the world is composed of overdeveloped nations and nations that will probably never develop much further, though we continue to use the euphemism "developing nations" in the public discussion to avoid insulting struggling societies. The fossil fuels won't be there for them any more than they'll be there for us. The capital for work-arounds will not be there either. Many so-called developing nations exist in geographically unfavorable places and are already deep into population overshoot. Think: Egypt. Now imagine Egypt with no more grain subsidies and only donkeys where they used to have trucks.

The trade-off for some of these poorer places is that everyday life there is still relatively primitive compared to life in, say, the Houston suburbs. Hyper-complexity itself is a comparative disadvantage when reality is sending you a strong message to simplify and downscale whatever you're doing. In short, people living at the core—the "developed" or "advanced" nations—could fall faster and harder than the people in the undeveloped places where life, however arduous, is already simple. So, it may be more accurate to view this as a leveling process. We'll know more in a few years.

The political landscape is shifting rapidly, especially among Western nations. The global banking system is groaning with stress after a ten-year artificial boom based on central bank interventions and unyielding engineered financial fraud. America's renewed status as the leading global oil producer is in jeopardy due to its overreliance on sketchy borrowing and poor net revenue flows among shale oil companies, from which all the recent, spectacular growth came. Global ecological damage is accelerating sharply. The effects of climate change—man-made or otherwise—have already reduced global food harvests while weather anomalies have become the norm. Reminder: things that can't go on, stop.

Chapter 10

CLIMATE CHANGE

I s it necessary to take a position one way or another as to how climate change is playing out? I'm not persuaded that it matters whether you are on board with the most authoritative science on the subject or if you're a "climate change denier," because I suspect the bottom line is that we're not going to do anything about it, and there may be nothing that realistically can be done. I tend to take the science seriously anyway. I can accept the likelihood that something momentous is underway, and that human activity in the industrial age is partly, perhaps wholly, to blame for it. I'm also willing to entertain arguments that there are other things going on, such as the sun-spot cycle, tectonic activity, and shifting planetary polarity, that are affecting Earth's climate dynamics. I don't pretend to know for sure. The arguments may be interesting, but they are probably academic.

The touchstone for climate change policy is the United Nation's periodic report under its Intergovernmental Panel on Climate Change (IPCC). The group receives data from scores of international scientific institutes and organizations. It would be too cynical to say that they are uniformly guided by hidden agendas. It's more a matter of consensual self-delusion. Facts are facts. Theories about facts are a level removed but still operate within the rules of science. About the best one can argue against them is that the models derived may not comport with the potential for apparently chaotic behavior resulting from deeply non-linear geophysical dynamics at work. Even supercomputers can fail to accurately predict the effect of that fabled butterfly flapping its wings in the Amazon jungle. Is it an Easter blizzard in

New England or a drought in New South Wales? The IPCC 2018 Report Summary states:

> Human activities are estimated to have caused approximately 1.0°C of global warming above pre-industrial levels, with a likely range of 0.8°C to 1.2°C. Global warming is likely to reach 1.5°C between 2030 and 2052 if it continues to increase at the current rate ...[43]

The models churn ahead from there. While it might be a mug's game to contest the conclusions, I would offer, at least, a vagrant theory of my own: The Fallacy of Exquisite Measurement, which states that *the ability to measure things down to very fine detail does not automatically confer a power to control things*. Which leads once again to the behavior I call techno-narcissism, an excessive faith in our technological prowess, especially as applied to rescue remedies from the woes of our time.

Paradoxically, the reams of prodigiously assembled exquisite data sets from around the world in the IPCC reports provoke a religious belief that surely something can and will be done about the climate trends so perfectly measured and modeled. The IPCC-backed 2015 Paris meetings that produced an international plan of action to reduce greenhouse emissions brought forth a program of impressively complex protocols, setting out what each country with its unique conditions will be required to do to ameliorate the effects of climate change—and theoretically save advanced civilization such as we know it. It was an admirable exercise in international cooperation that made a lot of people feel better for a while. And, of course, one of the first acts that President Donald Trump undertook was to pull the USA out of those agreements. That was widely regarded in the USA as a foolish move. I'd argue it was simply a recognition that the Paris Accord was a high-toned fraud. After all, governments are expected to do *something* in the face of emergencies, and the conclusions of the IPCC reports have persuaded the

---

43 "Global Warming of 1.5°C," Intergovernmental Panel on Climate Change, https://report.ipcc.ch/sr15/pdf/sr15_spm_final.pdf.

thinking classes in most nations to regard climate change as just such a major emergency (which it may indeed be).

August international conclaves that produce agreements based on credible science are a gold-standard type of *action*. Unfortunately, they lend themselves to pretense. Nations may claim they're determined to reduce greenhouse emissions, but the highly productive ones are not willing to sacrifice their fossil-fuel-based productivity. Leaders in some countries may have also concluded correctly that so-called sustainable energy or alternative energy, as currently developed, can't possibly take the place of fossil fuels at the current operating scale of industrial activities. The forgoing illustrates another principle of mine: Organizational Narcissism, *an excessive faith in the benefits of drawing up complex plans.*

As a basic matter, damage to Earth's biosphere that provokes changes in the atmosphere, oceans, and land masses is what I'd expect from any human population overshoot scenario in an ecology with limits—if not at this point, with nearly eight billion of us here, then at some point forward, sooner or later. Living organisms can't reproduce indefinitely without producing entropic consequences and degrading their habitat. The cliché is true: don't expect infinite growth on a finite planet. Have we reached that tipping point? Probably. Our waste products are self-evident. The amount of carbon dioxide (CO_2) in the atmosphere exceeded 400 ppm (parts per million) in an entire month for the first time as of March 2018. The Great Pacific Garbage Patch is a gigantic region of circulating plastic trash from just the past half century of manufacturing all that crap, and it is having a devastating effect on ocean wildlife. Plastic microparticles show up in the tissues of many organisms, and the larger stuff gets eaten and destroys animals' digestive systems, while grocery bags, discarded fishing nets, and beer halters strangle them to death. Microparticles are also showing up in American drinking water. Meanwhile, the new excess of CO_2 in the atmosphere is acidifying the oceans, harming and killing species, and upsetting deep relationships between them. All that is on top of the damage from sheer overfishing.

On the land, aquifers are being depleted and fouled at the same time. Water tables are dropping severely. Soils are being degraded, leached of vital food minerals, sterilized, and washed away. Tropical rain forests that play a crucial role in weather patterns are bulldozed and chainsawed to raise live-stock or extract ores. We've introduced untold thousands of synthetic chem-icals and pharmaceuticals into the ecosystem with some known effects and many unreckoned effects on all life-forms. The air quality in many cities is a foul soup of particulates and toxins. Superstorms have become the norm from Texas to Maine, and the western USA battles gigantic wildfires. All this is understood and documented.

China, with its enormous burden of 1.4 billion people, made an effort to limit population growth with Mao Zedong's one-child policy, established in 1979, but that is only one attempt at such an intervention on a meaningful scale. Mao was a despot, of course, and could decree impactful action with minimal opposition. After Mao's death, and the rise of Deng Xiaoping and his demi-capitalist successors, the extremely rapid industrial development of China seemed to validate the one-child policy, with the reward of a rising standard of living. The one-child policy was administered unevenly, more rigorously in the cities than among the rural peasantry. It had side effects that were predictable, such as infanticide. It was condemned by other nations as a human rights violation. And it was reformed in 2015, since the demo-graphics of elderly people to working-age citizens was deemed "unbalanced" by the party leadership. China has since moved to a two-child policy. For now, the Chinese people remain tractable, enjoying their new prosperity, while the government under Xi Jinping has created an Orwellian system of cyber "social control" to make sure the people mind what they do and say. This regimentation may not hold if the economy slips badly, as I believe it will, due to its energy quandary.

In any case, China's speedy march into modernity was accompanied at horrendous ecological cost to the country's air, water, and land. There are poor prospects of China voluntarily limiting industrial activities. Considering that is what pulled them out of the twelfth century in two generations, it seems to be about the last thing Chinese leadership would want to do. Of course, the leadership has often declared its intention to replace coal and oil with the usual cast of alt-energy methods: solar, wind, next-generation nuclear. But as I've suggested earlier, they will not be nearly enough to conduct the industrial economy at the current scale. Instead, it will eventually contract sharply. In fact, reports recently appeared saying that GDP growth in China projected for 2019 would be under 2 percent.[44] Who actually thought that growth rates of 8 or 9 percent registered in China in the early twenty-first century would continue indefinitely? My own view is that China's super-rapid industrial development was a manifestation of a global "crack-up boom," a one-shot stunt that would lead to an equally rapid, traumatic cessation and reversal of growth. As it happens, China could destabilize politically even faster than it became an industrial powerhouse.

No other nation has a population policy, including India, which has almost as many people as China. Europe is wringing its hands over a decline of its original national populations to birth rates below replacement level, while at the same time it struggles with an influx of refugees from the Middle East and Africa that has badly roiled the political scene. Nobody talks about population policy in the USA, and our fractured politics suggest that we're not even close to a public discussion about it—although the immigration furor is a kind of proxy for that. These days, even the system of government-subsidized family-planning clinics established in the 1960s is under assault. And, of course, Mr. Trump got elected on the promise to "*Make America great again*," which was widely understood to mean reindustrializing the country à la 1955 by bringing back all those factories we put on a slow boat to China. While there is loose talk about reindustrializing the US with so-called "green energy," the situation for alternatives to fossil fuels is the same for us as it is in China: it won't happen at scale or even close. Otherwise, we really have no intention to cut down on oil and gas use, and Mr. Trump wants to revive the coal industry as well. Plus, we haven't built a new nuclear power plant in going on thirty years.

Despite all the lofty talk, the leaders of the advanced nations really have no credible plans to do the one thing that is actually the most likely outcome, whether we like it or not: change our behavior by downscaling, relocalizing, and simplifying the terms of everyday life. Instead, techno-narcissistic

---

44 Patti Domm, "JP Morgan sees a slowdown coming, with economy growing at less than 2 percent in 2019," *CNBC*, November 20, 2018, https://www.cnbc.com/2018/11/20/jp-morgan-sees-a-slowdown-coming-with-economy-growing-at-less-than-2-percent-in-2019.html.

schemes for engineering Earth's weather and long-term climate pattern—"hacking the planet"—preoccupy the scientists who do all the measuring and modeling and write the reports, and their political counterparts who set the agendas.

These schemes include spraying aerosols and particulates into the various layers of the atmosphere to interrupt the journey of solar radiation to Earth's surface. One variation of this is called albedo modification. Albedo refers to the reflective capacity of a surface. Generally, white or light-colored surfaces reflect radiation back into space, while dark surfaces absorb it. Clouds and ice caps reflect radiation. Dark land masses absorb radiation and so do the deep-blue seas. The absorbed radiation expresses itself as heat. The interaction between warming oceans and land masses is self-reinforcing and destabilizing. The Marine Cloud Brightening Project at the University of Washington, Seattle, proposes to accomplish albedo modification by spraying seawater into clouds over the ocean, making them brighter and whiter to reflect radiation more efficiently. Cost? Hypothetical. Certainly not cheap. Probably requires airplane flights, which will pump more CO_2 into the atmosphere in the process. Recall the gag about the guy who tries to make his blanket longer by cutting a foot off the top and sewing it onto the bottom.

Another scheme is feeding iron particles to phytoplankton (microscopic plants) in the ocean, or "ocean fertilization." Theoretically, the iron promotes growth of the organisms, which take in CO_2 from ocean water they grow in. It is also hoped that the effect would move up the food chain and promote the growth of dwindling commercial fish stocks. Cost at the practical scale? Who knows?

A company in Switzerland called Climeworks proposes using huge fans to suck carbon out of the air with filters that attract and trap CO_2, and then put the carbon to use growing vegetables in adjacent gardens. They estimate that they'll suck about nine hundred tons of carbon out of the air per year, at a cost of about $600 per ton. It begs the question: What produces the electricity to run the fans? Solar panels? What about the sunk costs involved in fabricating the fans, the panels, and all the electrical management equipment, including hydrocarbons burned in the manufacturing process?

Another project proposes to build artificial undersea mountains to strategically shore up parts of the Antarctic ice sheets and keep them in place. If it involves large amounts of concrete, you can figure in the extremely high energy cost of manufacturing cement plus the carbon emissions entailed.

Another plan calls for dumping pulverized limestone into the ocean to "sweeten" the acidification underway. There's probably no shortage of limestone in the world, but consider the cost of pulverizing and transporting all that rock, and the size of the oceans.

And, of course, there is reforestation, planting trees, which is one scheme that is not necessarily techno-narcissistic and can be accomplished by human labor at different scales from the backyard on up. New England in particular has already seen a lot of reforestation by default since 1900 as abandoned pastureland goes back to woods, with no human intervention at all and at no financial cost (if you exclude the labor it cost in preindustrial times to clear the original first-growth forest).

My hypothesis is that only economic collapse will put an end to the industrial experiment and its destructive externalities at the current scale, though a cessation of fossil-fuel burning could worsen climate change effects if, as human extinctionist Guy McPherson says, the atmospheric aerosols produced by smokestack industry and car exhaust already reflect a lot of radiation that would otherwise reach Earth's surface and heat it up—a depressing possibility. Ecological degradation and economic collapse are just two sides of the same coin, since *economy* is the way we manage our human *ecology*, and mismanagement of one affects the other. They will change in tandem.

The weak link in our economy is the feature we call finance, which has been failing in one way or another since the dot-com bust. It slipped very badly again in 2008 and was temporarily salvaged by nations borrowing massively from the future to cover current expenses and to bail out insolvent "systemically important institutions," i.e., Too-Big-to-Fail banks. The problem, which ought to be obvious by now, is that the future will not provide the amount of collateral that was pledged as security for all that debt—namely, "growth" of ever-more industrial activity and the surplus wealth it would hypothetically produce. When old debts are seen as unpayable, you will not generate new debt. The game stops. The global manufacturing and supply lines stop, too. You have to think of a new way to live.

These days, it comes down to a question of adaptation versus mitigation. Adaptation will be very difficult as many advanced systems we depend upon stop operating and then reinforce each other's failures. But adaptation to new and different conditions of daily life at least points the way toward practical strategies for carrying on, for instance, learning to live at a much lower scale of economy, with all that implies.

Our attempts to mitigate the quandary of running a banking system based on limitless growth on a finite planet have only introduced more pernicious (and dishonest!) complexity to a foundering system. Attempts to mitigate climate change won't keep business-as-usual going either. Anyway, the reality of our financial weakness only suggests we don't have the capital resources (a.k.a. the "mojo") to make all that dreamed-of mitigation happen, even if it was flawlessly designed. So, realistically, except for a few lame gestures, adaptation is all we've got. For individuals and families, it means having to take responsibility for making the right decisions. The confusion at even the most refined layers of the intellectual class is so great now, so vitiated by wishful thinking, public relations bullshit, and political opportunism, that the responsibility weighs extra heavily on individuals to think for themselves through the fog of yammer.

Adaptation may be more difficult this time than in earlier instances of historical turmoil. Climate change—whether it means fire or ice—is coming simultaneously with the end of the fossil-fuel era. The hyper-complexity of our current arrangements is so extreme, the scale of our vital activities so great, and rapid growth of human population so unprecedented, that we have a long way to fall in any plausible reset of civilization. As it happens, I'm not a human extinctionist. I believe the human project will continue—though nuclear war and other gross catastrophes can't be ruled out. Assuming humanity navigates through these straits, we may be lucky to land in the equivalent of a medieval standard of living when we stop falling. For now, the question comes down to heuristics. Put the picture together the best you can. In the years ahead, thinking for yourself will be the rule, so get used to it.

Chapter 11

THE FOOD QUESTION AND OTHER NAGGING DETAILS

Assessment of many studies covering a wide range of regions and crops shows that negative impacts of climate change on crop yields have been more common than positive impacts (high confidence).

—The IPCC report[45]

By "high confidence," this assessment by the United Nations Intergovernmental Panel on Climate Change means a high-percentage likelihood. Whatever one might learn about global food production and its future, keep in mind that in the current moment of history, most of the food grown around the world, and certainly in the "developed" countries, is sown, cultivated, and harvested with the help of fossil fuels, and most of that "agri-business" style of farming is done at the very large scale. So, thinking about future food production, you must figure on a problem at least as significant as an unstable climate: the end of affordable oil and gas and all their by-products, especially pesticides, herbicides, fertilizers, and motor

---

45 "Climate Change 2014 Synthesis Report Summary for Policymakers," Intergovernmental Panel on Climate Change, https://www.ipcc.ch/site/assets/uploads/2018/02/AR5_SYR_FINAL_SPM.pdf.

fuels. By and large, the models produced by esteemed scientific institutes about future food production leave that part out. Many seem to assume that the current practices we use now will just continue at the giant scale, and only the climate will change—and we'll just see what happens. Or that less technology-dependent methods will be scaled up to feed a global population estimated to reach ten billion later this century. This allows them to avoid discussion of a catastrophic crash in food production (and of population). It's pretty crazy and hard to fathom how such deeply educated, experienced empiricists can overlook it.

The world went through a preview of coming attractions for food scarcity in 2006 to 2008. Oil prices took off and reached an all-time high of $147 a barrel in July of 2008. That event affected the price of fossil-fuel-based agricultural "inputs," but there were other factors. One was the "improved" diet of the rising populations of the so-called developing world. They were eating more meat, which is a more "resource-intensive food." (Raising one kilogram of beef takes seven kilograms of grain.) Another factor was the ramping up of ethanol production as a motor fuel. A lot of corn was diverted to that project in the US, and sugarcane in Brazil. Yet another factor was the monkey business in commodities-markets futures trading during a period of rampant speculation and banking irregularities that would terminate in the Great Financial Crisis during the harvest season of 2008. A series of weather events also aggravated the problem: drought in Australia's most productive wheat region, unseasonable rains in India's Kerala grain belt, a cyclone in Southeast Asia that damaged the rice crop, a heat wave in California's San Joaquin Valley. Apart from weather, a stem-rust disease in the Middle East and North Africa dropped yields, with Yemen hit especially hard (and in *failed state* chaos a decade later). Prices of basic foods rose. Predictably, political unrest followed, with riots and protests in many countries. The parade of mutually reinforcing systems failures between weather, finance, oil, and politics was impressive.

One planting season doesn't necessarily make a trend, but let's look at some more recent news and figures from the year 2018:

Unusual weather affected European crops significantly. A heat wave in northern Europe resulted in the smallest harvest in six years among Germany,

Poland, Ireland, the Baltic states, Sweden, the Netherlands, Denmark, and the UK. It was the UK's hottest summer in four decades. Sweden got only 12 percent of its usual rainfall between May and August, and wildfires broke out there. In contrast, France, Europe's leading wheat producer and a major exporter, suffered extended soaking rains. German and Polish wheat harvests were down over 6 percent, and German farmers had to import grain from Romania to feed their livestock.

Wheat production in the US and Canada was up nearly 4 percent in 2018, but Russia, the world's number-two exporter, was down over 17 percent, and Ukraine, another breadbasket, was down over 7 percent. China's production fell a slight 1.3 percent. South African wheat production fell 8 percent under severe drought conditions.[46] Parts of the wheat-growing regions in New South Wales, Australia, the world's fourth-leading wheat exporter, saw the lowest rainfall levels ever recorded, and University of Melbourne scientists studying climate patterns noted a steady decline in rainfall going back thirty years. The 2018 wheat yield fell over 20 percent to a ten-year low. Sydney and Melbourne, the country's two largest metro areas, were also hard-hit by drought as well. And so it went. Not a worldwide catastrophe, yet, but trending down.

Another thing trending down in 2018 was global oil prices, due largely to the US shale oil production stunt, a by-product of artificially ultra-low interest rates. West Texas Intermediate, the US benchmark, fell steeply from a high of $75 a barrel in September to $45 a barrel late in the year as US production soared way past its previous peak back in 1970. Brent crude (the European benchmark price) followed suit in price. Oil could easily rise back above the 2018 high, especially if interest rates rise and shale oil producers can't roll over their old loans to keep drilling and fracking. Volatility in the oil market is a killer for farmers, who need to carefully rationalize their "input" costs from planting to harvest.

The National Academy of Sciences report issued in midyear 2018 offered a series of predictions and scenarios. Corn (called maize outside the US) will

---

46 "World Agricultural Production," US Department of Agriculture, Circular Series WAP 8-19, August 2019, https://apps.fas.usda.gov/psdonline/circulars/production.pdf.

be deeply affected by climate-warming trends because just a few countries dominate production and trade. The top four corn exporters account for 87 percent of total corn exports. Around the world, corn is the leading livestock feed. The report predicts more frequent "synchronous price shocks."[47] Crops have an optimal performance window within a temperature range. Heat stress reduces harvests. "With continued warming under business-as-usual greenhouse gas emissions, global crop yields are expected to decline significantly," the report states. "For every degree increase in global mean temperature, yields are projected to decrease, on average, by 7.4% for maize, 6.0% for wheat, 3.2% for rice, and 3.1% for soybeans." That would likely raise prices as there will be ever more mouths to feed around the world—for a while anyway. The report overlooks how rising commodity prices would be amplified by distress in financial markets due to debt saturation and bond default. Highly rationalized agri-biz-style farming can't tolerate highly volatile commodity prices, especially since borrowed capital is one of the major "inputs" after fertilizer and the rest. Changes in yields of more than 40 percent are predicted in the United States, Mexico, eastern Europe, and southern Africa. The smoothly operating global trade in food of recent decades that has managed to feed the growing world population would be deeply disrupted. The report also tends to overlook the coincident likelihood of geopolitical disorder as that picture evolves. During the 2006–2008 crisis, large corn-exporting countries, including Brazil, Argentina, and Ukraine, imposed export bans. Before too long, it could be every country for herself, or every province, or county, or village.

So, what can humanity do about the coming food crisis? Some of the answers are implicit in the portraits of early adapters in Part Two of this book: grow food by methods other than high-input agri-biz. Do it intelligently, using new knowledge developed in recent decades combined with proven traditional practices, and do it at a scale that comports with the resource realities of the future. Mark Shepard's model of silviculture out in

---

47 Michelle Tigchelaar, David S. Battisti, Rosamond L. Naylor, and Deepak K. Ray, "Future warming increases probability of globally synchronized maize production shocks," *Proceedings of the National Academy of Sciences* (PNAS), June 26, 2018, https://www.pnas.org /content/115/26/6644.

the Driftless Area of Wisconsin is one model. Kempton Randolf's farm and distillery is another.

The Drawdown Project, a coalition of geologists, engineers, agronomists, researchers, fellows, writers, climatologists, biologists, botanists, economists, financial analysts, architects, companies, agencies, NGOs, and activists, is an exemplary, comprehensive, organized endeavor. Considering its eminent membership, the defects in the group's collective thinking are impressive. The Drawdown Project proposes to overcome the problems of climate change—actually reverse it—and feed the world by changing just about every system for everyday life currently in place, along with a broad agenda of farming reforms under the rubric of *regenerative agriculture*.[48] It is an extremely ambitious project, presented with elegant rationality and concision, and it includes many innovations in systems redesign that have been well publicized in recent times, since the public awareness of global disruption got traction in the twenty-first century. They cover all the bases from alternative energy to transportation, recycling, the electric grid, walkable cities, managed grazing, educating women, biochar, wetland management, and dozens of other categories.

And yet the Drawdown Project makes the same mistakes as the National Academy of Sciences proposals: techno-narcissism and organizational grandiosity. It doesn't acknowledge anywhere in its manifesto that the scale of human activities must necessarily be reduced. It appears to assume that we can continue most business as usual by other means. For example, their pitch for electric vehicles is that they reduce CO_2 emissions "by 50 percent if an EV's power comes off the conventional grid [and] if powered by solar energy, carbon dioxide emissions fall by 95 percent."[49] Really? What is the embedded energy cost to manufacture all those solar panels? Can they even be fabricated at scale without fossil fuels? Did they calculate how much silver would be required? I also suspect they did not figure in the embedded energy costs in manufacturing an EV, including the mining operations, smelting ores, the

---

48 Project Drawdown, https://www.drawdown.org.

49 "Transport Electric Vehicles," Project Drawdown, https://www.drawdown.org/solutions/transport/electric-vehicles.

need for sophisticated plastics, rare-earth minerals for electric motors, and long manufacturing supply lines.

More to the point: Is the motive behind EVs based on a wish to continue living a Happy Motoring life in suburbia? When do we recognize that suburbia is an obsolete and profligate settlement pattern that has many more problems than just the issue of getting around its vast distances? Also, how does the EV scheme pencil out in a society that no longer has the capital for *x* millions of car loans, or a shattered economy that has left a huge portion of the middle class unworthy of credit. I'm not persuaded that the public will be able to afford using them as ever-circulating driverless taxicabs if they are not getting regular paychecks. Frankly, I don't see EVs playing much of a role, even in the transition between the hopped-up fossil-fuel economy and the austere economy to come—since the transition will necessarily be the most disorderly part of the phase change.

Likewise, the Drawdown Project's riff on airplanes of the future:

A century after the first commercial flight, the aviation industry has become a fixture of global transport . . . and of global emissions. Today, some 20,000 airplanes are in service around the world, producing at minimum 2.5 percent of annual emissions. With upwards of 50,000 planes expected to take to the skies by 2040 . . .[50]

Say what? I'm much more inclined to think that there will be little left of commercial aviation in 2040, not that it will be two and a half times bigger than it is now. Their argument assumes that aviation fuel—basically unleaded kerosene—will continue to be available in the same (or greater!) volumes than today. It assumes a lot more, too, such as the continuation of tourism at present rates, of business travel connected with global corporate activity, the sustainability of mega-cities and metroplexes, with their expensive airport infrastructures, and of geopolitical stability in a world of failing states. I'm sure it is a cool and fun exercise to imagine evermore elegant technologies,

---

50 "Transport Airplanes," Project Drawdown, https://www.drawdown.org/solutions/transport/airplanes.

but that ignores the central problem of making further overinvestments in technological complexity as the already-existing hyper-complexity groans with diminishing returns. Extreme global interdependency means it takes fewer failures to send cascading failure thundering through the system.

Alice Friedemann, the Berkeley, California, technology writer, former banking executive, and author of *When Trucks Stop Running*, has made the point repeatedly that long-distance freight trucks can't run on electric engines, and that there really is no substitute for diesel fuel in that industry.[51] The batteries would have to take up most of the cargo space and weight. Light-duty short-haul trucks performed poorly on lithium-ion batteries when road tested by two national companies, Frito-Lay and Staples, and cost about three times as much as the internal combustion versions. There is probably not enough lithium available to outfit the world trucking fleet in any case. No batteries are compact and powerful enough for farming vehicles (tractors, harvesters) scaled to the current practices of agri-biz, or the giant earthmoving machines used in mining.

"[E]ssential supply chains depend on trucks nearly completely," she writes. They even deliver the diesel fuel that the trucks run on, as well as other essentials, such as the chemicals used for most municipal water-treatment plants and the food that most Americans eat. By the way, forget about hydrogen as a motor fuel. Hydrogen has been demonstrated to be a net energy loser under any scheme currently known. (Go back to *The Long Emergency* where that was discussed in detail.)

These cautionary notes suggest that we face systemic breakdown sooner rather than later, and that when it occurs, advanced societies will be overwhelmed. Our systems may not recover. The just-in-time economy with all its networked global elegance means that there is always just enough stock of anything to conduct business—and no more than just enough. Which means that when the transport stops, pretty soon everything stops. The popular adage states that "we are three days away from anarchy," based on the fact that supermarkets around the USA contain about three days of food supplies.

51 Alice Friedemann, *When Trucks Stop Running*, New York: Springer Books, 2015.

The thinking displayed in the Drawdown Project's manifesto ignores a primary reality of our predicament: that increased complexity leads to increased risk of systemic breakdown. It's as if they can't imagine a world without a continuing expansion of human activities, as represented in economic growth. That is their only context. All that needs to be done is to "green up" the growth. Contraction changes everything. When such a large cohort of trained scientists overlooks this, you have to conclude that we're in serious trouble.

Chapter 12

EXTINCTIONS NEAR AND FAR

This will be a short chapter. The material is depressing even to me, and I have been keenly aware of the insults to the planet manifesting in my lifetime.

A so-called Sixth Extinction appears to be underway. When I see tuna on the menu in a restaurant, my heart sinks a little. How many of these magnificent fish are left in the oceans, and why do we continue to catch them? (Rhetorical question.) We catch them because that's what humans do, especially humans who are used to catching them and eating them from time immemorial, and we will continue to do so until there are no more left. That's how we roll. That kind of behavior has been hugely exacerbated by the high-tech commercial rationalization of "harvesting" ocean catches and selling them into mass markets. Until I was a teenager, I probably wasn't fully conscious that "Chicken of the Sea" was even a fish—even though the item was called a "tuna fish sandwich." It was just something that came out of a can, which you mixed with mayonnaise, and constituted a tasty sandwich spread. But that is exactly how our kind of high-tech economy abstracted us from reality. The fact is, 97 percent of the bluefin tuna that were in the oceans before industrial-scaled, mechanized fishing began are gone now.[52]

I take some consolation in the thought that a long emergency will begin to stop this process as the human module of life on Earth contracts and

---

52 Brooke Jarvis, "The Insect Apocalypse Is Here," *New York Times*, November 27, 2018, https://www.nytimes.com/2018/11/27/magazine/insect-apocalypse.html.

resets to a much lower level of activity. But the damage to the planet's diverse ecologies is profound, and surely there will be a difficult recovery for all the other species here with us. Some may never recover, and Earth will move into a new epoch where many empty niches beg to be filled by new life. Human complexity has revved up at the expense of complexity in the rest of the natural world. Science has learned a lot about the interdependencies of species and their environments, but there are still vast realms of mystery about the richness of these interactions—and the unexpected consequences of interrupting them. The *New York Times* article cited makes a useful distinction between extinction of a species (all gone) and functional extinction (still there, but not in enough numbers to play its necessary role). Entropy kicks into high gear and an integrated natural system tilts into disorder. The knock-on effects are chastening, as the article's author, Brooke Jarvis, reveals:

A 2013 paper in *Nature*, which modeled both natural and computer-generated food webs, suggested that a loss of even 30 percent of a species' abundance can be so destabilizing that other species start going fully, numerically extinct—in fact, 80 percent of the time it was a secondarily affected creature that was the first to disappear. A famous real-world example of this type of cascade concerns sea otters. When they were nearly wiped out in the northern Pacific, their prey, sea urchins, ballooned in number and decimated kelp forests, turning a rich environment into a barren one and also possibly contributing to numerical extinctions, notably of the Steller's sea cow.[53]

"Trophic cascade" is the term for these damaged relationships, resulting in barren and impoverished ecosystems. Trapped in our human ecosystem with all its artifice, we seem to barely notice until reality whaps us upside the head with the proverbial two-by-four. And by then it is too late.

The so-called "insect apocalypse" is another feature of the extinction trend. Over the past thirty-five years, the abundance of invertebrates such as

---

53 Ibid.

beetles and bees has decreased by 45 percent.[54] The monarch butterfly census is down 80 percent. Scientists in Britain report that monitored insect species are down by 45 percent, with their ranges also reduced. Studies out of Germany are even more discouraging. Some insect census experiments showed populations 80 percent lower than in 1989. Insect populations were shockingly lower in the pristine Luquillo rain forest of Puerto Rico (also known as El Yunque National Forest, America's only national park with a rain forest), where scientists have been studying bugs for decades. Sadly, 98 percent of ground insects were gone, and 80 percent of the bugs in the tree canopy were missing. Pesticides are not even a factor there. (In the farming regions of Puerto Rico, pesticide use has fallen by 80 percent since 1970.) "Everything is dropping," said Bradford Lister, a biologist at Rensselaer Polytechnic Institute, who began studying the rain forest in the 1970s. Something else is up, possibly that even in the steamy tropics, small changes upward in temperature patterns affect the ability of insects to reproduce. Since the 1970s, the average high temperature in the Luquillo forest increased by four degrees Fahrenheit.

Collapsing bee colonies have certainly been noticed in the USA, where the pesticide family of neonicotinoids is one suspected culprit. These popular, "innovative" chemicals, which are coated onto seeds or sprayed onto young plants, were designed to persist in the landscape for a long time so farmers would not have to re-spray. They are neurotoxins (nerve agents), and a clue to the puzzle of "hive collapse" is that the hives are often found to be empty of bees altogether, even dead ones. This suggests that the bees' sensitive orientation and navigation systems are harmed and they can't find their way home. Neonicotinoids have been detected in streams, honey, garden flowers, and wildflowers. These chemicals represent only one of many stresses on insect life induced by human activity. Global trade spread the Varroa mite from Asia to other parts of the world, arriving in the USA in 1987. It has been deeply damaging to bee colonies, along with the emerging new diseases Israeli acute paralysis (a virus) and the gut parasite Nosema.

---

54 Ben Guarino, "'Hyperalarming' Study Shows Massive Insect Loss," *Washington Post*, October 15, 2018, https://www.washingtonpost.com/science/2018/10/15/hyperalarming -study-shows-massive-insect-loss/?utm_term=.a9918e0b3511.

If the insect populations are falling, the effect moves up the food chain to the animals that depend on them for food. Lizard and frog populations are in decline. Half of all birds in Europe's farming landscape have disappeared since the 1980s. Nightingales and doves are down 50 and 80 percent respectively. Partridges are down 80 percent; 44 percent of 460 bird species in Canada show falling population trends since 1970, and forty-two species common to the United States, Canada, and Mexico have lost 50 percent or more of their populations since the 1980s.[55] In Canada and the northeastern United States, the total reduction in insect-eating birds between 1970 and 2010 was greater than 60 percent. Acid rain was a suspected culprit in the 1970s, but that factor has abated with smokestack scrubbers. The rise of ocean temperatures affects the food supplies of marine birds; the Antarctic king penguin colony shrank by 88 percent in thirty-five years.

A thousand animal species have gone extinct in the past five hundred years: the dodo, of course; passenger pigeons; woodland bison; Puerto Rican parrots; and ivory-billed woodpeckers are just some of the better-known names. Conservation scientist David Wilcove estimates that there are altogether 14,000 to 35,000 endangered species (of all plants and animals) in the United States, which is 7 to 18 percent of US flora and fauna.[56] Worldwide, 16,928 species are threatened. Amphibians (frogs, toads, and salamanders) are especially sensitive to environmental degradation and are fast disappearing. Coral reefs are "bleaching out" and dying. Warming oceans are suspected in killing the algae that live symbiotically with the coral-building organisms and supply them with their food. The huge variety of fish, mollusks, and other organisms associated with the life of the reefs are affected, too.

The death of the world's oceans is perhaps the most dispiriting part of the story, which is often described as a combined tragedy of climate change and overfishing by rising populations of hungry humans patrolling the seas with commercial fishing systems that sweep whole regions clean of life. The

---

55 Bradford C. Lister and Andres Garcia, "Climate-driven Declines in Arthropod Abundance Restructure a Rainforest Food Web," *Proceedings of the National Academy of Sciences* (PNAS), October 30, 2018, https://www.pnas.org/content/115/44/E10397.

56 "The Extinction Crisis," Center for Biological Diversity, https://www.biologicaldiversity .org/programs/biodiversity/elements_of_biodiversity/extinction_crisis/.

United Nations' spinoff group, the Integrated Regional Information Network (IRIN), estimates that "the oceans have absorbed 93 percent of the heat generated by human activity since the 1970s."[57] Ocean warming and overfishing threaten 70 percent of fish populations. If current trends persist, there will be no fish left by the year 2050. For the human population, the problem is especially acute in the lower latitudes, where people in the poor, coastal, tropical regions depend on wild fish for basic nutrition. A study by the Global Ocean Oxygen Network under the UN's Intergovernmental Oceanographic Commission (IOC) reported that oxygen concentrations have declined in "large swaths" of the oceans, promoting "dead zones" slowly suffocating the life there. Pollution—nutrients from agriculture, sewage, and industrial waste—is exacerbating the damage caused by temperature rise. These, plus rising sea levels and ocean acidification, threaten the mangrove swamps, estuaries, tidal marshes, and other "nurseries" where fish and mollusks spawn.

Fish can at least migrate elsewhere if their local habitat is degraded. Under the most plausible climate scenarios, species are expected to move from the tropics to higher latitudes, but it is much harder to predict the effect of these migrations on ecosystems as a whole. It is liable to create additional disruptions.

During the two hundred years of the Industrial Revolution, the rising CO_2 level in the atmosphere has expressed itself as increasing acidity of the ocean, which absorbs about 30 percent of this released CO_2. As this happens, a series of chemical reactions raises the concentration of hydrogen ions in the ocean. It affects available carbonate ions, which many of the ocean's invertebrate organisms use to construct their shells of calcium carbonate. It also affects sea urchins, corals, and planktons. The tiny animals at the bottom of the food chain, which are eaten by everything from krill to whales, depend on carbonate structures that dissolve in acidified ocean water. "Researchers have already discovered severe levels of pteropod shell dissolution in the Southern Ocean, which encircles Antarctica," states a report by the National

---

57 Jared Ferrie, "A Perfect Storm: Climate Change and Overfishing," *The New Humanitarian*, September 19, 2016, http://www.thenewhumanitarian.org/feature/2016/09/19/perfect-storm-climate-change-and-overfishing.

Oceanic and Atmospheric Administration (NOAA).[58] Larger animals without shells are affected as well. Pollock, an important commercial fish, loses its ability to detect predators, and clownfish lose the chemo-sensory ability to identify their habitats among sea anemones. These are only the effects we know about.

Changing the ocean's basic chemistry is a dangerous game. It's been thirty-five million years since Earth last reached the atmospheric CO_2 levels that are being recorded today. There are several earlier instances, too, recorded in undersea sediment cores. But the rate of change has never been as fast as the current situation. Even in periods when the CO_2 levels were higher, the *rate of change* was slower, and the oceans had time to buffer the effects. That's not the case now. It's hard to conceive an outcome in which we can have dead oceans and a still-living Earth.

Obviously, this is not a comprehensive discussion of all the disturbances to Earth's ecology, and it is not intended to be. Rather, it's a casual survey of major trends that are pretty daunting. The reader is probably asking now: Okay, can you please tell us what we can do about these grim trends? Can international cooperation be organized to arrest the range of insults to the planet by humanity? Can we not marshal science and technology to take positive action?

I've already described my objections to techno-narcissism and organizational grandiosity. We think too highly of our magical abilities to control the things we are so busy measuring. What will finally change the picture is the economic collapse of techno-industrial society. It will compel the cessation of many destructive activities and the slowing and diminishment of many more. Even after collapse, many of the disturbances and vicious feedbacks of the present time will continue to play out and worsen, while humans do less to aggravate the situation—and there will be far fewer of us around to do it. We've set too many unpredictable dynamics in motion to feel smug about being able to work around them.

---

58 "Ocean Acidification," National Oceanic and Atmospheric Administration, https://www.noaa.gov/resource-collections/ocean-acidification.

That verdict is not an invitation to despair and passivity. It merely establishes some important parameters for what you can do, and what's worth doing. Grandiose international covenants are a waste of time. Carefully picking a favorable place to live is a good investment of your time. Where's it going to be? Orlando, Florida? The hills of Vermont? Bad Axe, Michigan? Better study up. Doing what you can to live locally in expectation of major economic systems failure is a good idea. Working to downscale, simplify, and arrange your life in accord with real circumstances, and preparing to occupy a valued role in a community at the small scale is within your power to accomplish. Quit wringing your hands and get on with it. In the next chapter we'll look at the vectors of systems collapse.

Chapter 13

MONEY, OIL, AND THEIR BY-PRODUCTS

The crash of 2008 and the Great Financial Crisis it spawned was the beginning of the end. In the event, the developed nations used the last of their creditworthiness to paper over the crisis and string out the reckoning for another decade. That creditworthiness was based on the assumption that the nations could generate enough economic growth in a dazzling postindustrial techno-utopian future to enable them to eventually cover their borrowings. The upward progress of the human project seemed irresistible, even with periodic setbacks such as world wars and depressions. Global economies, it was thought, would always return to the growth trend line, and that growth could be considered the collateral for ever-more borrowing. Growth meant more, more of everything, ever-more *more*.

Alas, that assumption is false. Despite the brilliant pixel display on the iPhone 8 and all the chatter about robots, the coolness of Amazon's "virtual assistant" Alexa, and Elon Musk's adventures in rocketry and self-driving electric cars—despite all that magic, in 2019 the global economy was manifestly contracting, not growing, and all the signs were that the contraction was apt to be persistent, possibly even permanent, as far ahead as anyone could see. And so, the realization began to spread through banking and finance that without growth, the developed nations were actually no longer creditworthy and, more to the point, that the public would soon catch on. That is, if expected future growth was collateral for all these massive borrowings—about $240 trillion globally as I write—and there was no more growth, then the debt would never be repaid, and the nations in question would probably not be able to borrow anymore, meaning further, no more bid on bond sales, unless the central banks bought all the bonds themselves. And, if they did, it would be an arrant fraud, and the attempt to do so would have nasty consequences, most likely a currency crisis in which the nation's money itself lost its value, as people lost faith in the operations that supported it.

As I write, the Federal Reserve has decided to halt its brief "tightening" experiment of raising interest rates, and therefore restricting credit. In July of 2019, it resorted to lowering the basic interest rates for the first time since 2008. The Fed's tacit aim the past decade has been to support the share price of stocks. This came from the one-dimensional idea that higher stock prices mean the economy is doing well. It was, in fact, a perversion of the relationship between finance and the actual work of creating wealth in a real economy. The ten-year-long regime of "easy" credit and ultra-low interest-rate policy induced corporations to borrow money cheaply to buy back their own stock (from other shareholders), thereby boosting the share value by removing total outstanding shares on the market.

The main result was that jacked-up share prices tended to reward top executives with large year-end bonuses, and more valuable stock holdings, regardless of the performance of the company. The bonuses were usually granted by boards of directors who were appointed by the same people getting the bonuses. These machinations were just another form of classic and massive malinvestment. Before ultra-low interest rates became the new norm, corporations generally borrowed money in order to invest in their own future production—to build new factories, buy new equipment, hire new people, and fund research and development into new and improved products. Borrowing strictly to fatten executives' pay is not legitimate investment. It's a form of racketeering.

The Fed's tightening program was ostensibly undertaken to "normalize" interest rates, that is, to correct the warped incentives of abnormally low rates and return to interest levels that realistically price the time value of borrowed money. *Normal* interest rates of, say, 4 to 8 percent reward savers—if inflation is running at less than that, which is also ostensibly normal. Ultra-low interest rates make saving pointless, especially if the "real" rate is negative, which is what happens if inflation is higher than the interest rate. For instance, if you put money in a savings account that pays 1.75 percent annual interest, and the inflation rate is 2.25 percent, then you are a net annual loser. If there's no incentive to save, one result will be impaired capital formation, the accumulation of surplus wealth that can be used to generate new wealth-producing enterprise. You can also view it as mismanagement of all the work that people in society do to produce things of value.

Many savers are traditionally older people who seek a low-risk but regular return on their savings. Removing the incentive for saving either pushes them into higher-risk investments like stocks, or prompts them to just spend down the money before it loses even more value. Pension funds, too, are traditionally inclined to seek annual income from low-risk interest-bearing securities, and they need to show reliable gains year-in and year-out in order to pay their obligations to retirees and stay solvent. Pension funds have been brought close to ruin by the ten-year experiment in super-low interest rates. And, having been shoved into higher-risk investments—stocks and higher-yielding "junk" bonds—they run the risk of getting wiped out altogether in a stock market crash or a junk bond default.

Normalizing interest rates would seem to be the logical policy. However, the government itself has racked up so much national debt—$22 trillion as I write—and the annual interest is so enormous that it can't take higher interest rates. Interest payment on US debt for the first three months of fiscal 2019 was $164 billion, 12 percent more than the same period the previous year. As interest rates rise, US treasury bonds, bills, and notes have to be "rolled over" at higher interest rates, and US taxpayers are on the hook for payments that take more and more of the federal budget.

The Federal Reserve is between a rock and a hard place. The super-low interest rate policies have perverted the "price discovery" process in markets and many other fundamental functions of finance, but returning to normal rates will bankrupt the government. They're damned either way. So, now what . . . ?

Continuing to pile up more debt at ultra-low interest rates will lead to some form of default or refutation of debt. Bonds and loans will not be repaid because they can't be. The nation will be bankrupt, unable to continue borrowing, and unable to pay its own ongoing costs of operation. When debts are defaulted on in a system where money itself represents debt, there ends up being a lot less money in the system. Nonpayment of debts extinguishes debt-based money. The public goes broke along with business and government. The further result is a crashing standard of living. People go hungry, lose their jobs, their homes, their sense of reality. It's an excellent recipe for political turmoil, and that's exactly what we've been seeing.

The Federal Reserve can resort once again to the failed policies of ultra-low interest rates and buying the bonded debt of the government—lending it lots more money. That is, the Fed can resume so-called "Quantitative Easing" (QE) and "print" money. There will be a lack of bond buyers (lenders) in whatever is left of the regular bond market because it will be obvious by then that the government is a deadbeat—that the only way it can pay back the bondholders is in money that has less value than it did when it was borrowed, because the government is repaying old debt in new borrowings that it has, in effect lent to itself. So, it completes the circle jerk of borrowing from its own central bank in order to pay back the central bank. Since finance operates within the laws of physics, entropy steps into this circle-jerk process, and the equivalent of entropic heat loss in physics is value loss in money. Thus, under the fakery of "financialization," the new money that is lent into existence is *money with its value removed.* That is the essence of monetary inflation. Remember: having plenty of money that has no value is just another way of being broke.

The first three iterations of QE produced only feeble, nominal economic "growth," and much of that was derived from systemic national rackets: war profiteering; grossly inflated medical payments; Federal mortgage guarantee shenanigans with inflated house prices; college loan huckstering; Wall Street's fees, commissions, and swindles; pervasive accounting fraud; and lots more. Rackets, of course, are dishonest or illegal ways of getting money. Financialization itself, at the grand scale, was a racket—substituting swindles and frauds for the old economy of industrial production.

Nothing lasts forever, of course, and wonderful as it was, the industrial economy couldn't last forever. It produced too many destructive externalities, including global population overshoot, resource depletion, and planetary ecological impoverishment. So, as it became apparent that the game couldn't go on, a lot of the smarter folk in society turned to racketeering to keep making money at all costs—even doctors and college presidents! The really magical part was that they managed to make it socially normative and ethically acceptable, even while the middle class was systematically impoverished by their rackets—so that in the year 2019, any poor schnook who took his child to an emergency room with appendicitis could find a bill for $97,000 in his mailbox two months later, with no plausible explanation for the charges—and absolutely nobody from the hospital billing director to the state attorney general would do anything to correct the injustice, or even listen to his complaint.

The reason for the gross dysfunction in American money matters was a fatal combination of overinvestment in complexity, imperial overreach, ethical and moral decay, and the civilizational life cycle in history that tends to run about 250 years, which is pretty much where the USA finds itself right now. But as much as any of that, the technical reason for the end of growth in our mature industrial economy was the economics of our fossil-fuel energy supply, oil in particular.

In Part One, I described the relationship between low-interest-rate policy and the shale oil "miracle." A year has gone by in the writing of this book, and an update in that situation is in order.

Just to review: the shale oil boom has been an impressive stunt made possible by super-low interest rate, cheap financing. The catch was that most of the companies engaged in shale oil production lost money doing it and borrowed ever-more money to boost production in the hope that higher volumes of oil production would somehow allow them to service their debts, continue operating, and eventually turn a profit. This has proven to be a disappointment—and with tremendous irony, because even under these unfavorable circumstances, the shale oil boom produced so much oil in 2018 that it not only surpassed the previous all-time production peak of about ten million barrels a day from 1970, but soared majestically above twelve million barrels a day, making the US the world's leading oil producer again. No wonder the public is confused.

The production boom also happened to drive down the price of oil substantially, making it even more difficult for the shale oil producers to show a profit. Across 2018, benchmark West Texas Intermediate crude (WTI) started the year at about $60 a barrel, climbed to $76 a barrel in October, and sank back to $42 a barrel in December. In early 2019 it was back up a bit at $60 a barrel and by late summer it was $55. Permian shale oil was selling for less than WTI because, as discussed in Part One, it is a very different combination of distillates that has less value than conventional crude. Most of the shale producers were unable to show a profit in the first two quarters of the year, and the investment community was finally beginning to shun them.

The bottom line of all this is not so complicated: oil is the primary resource of our industrial production. If much of the USA's oil is basically uneconomical to get out of the ground, then industrial production and all its ancillary activities can't go on as before, and there won't be *growth*. Without growth, the economy collapses. The first symptoms of distress show up in finance, banking, and the cost of money, which is a proxy for all that. And if the financial sector gets into trouble, shale oil producers surely will not get the loans they need to keep pulling the oil out at no profit, a grim feedback loop.

So, it is a mistake to read the current statistics on shale oil production and draw the conclusion that all is well. My own conclusion is that shale oil will be a shockingly short-lived "miracle," and that the lending gravy train for it will go off the rails. Because the depletion rate of shale oil wells is so steep, the industry can't function without high rates of continual capital spending, and they can't spend without borrowing. The wells have to be drilled incessantly and fracked (with countless truck trips hauling fracking sand, water, and chemicals to the wells). The industry is caught in the Red Queen Syndrome: running as fast as it can to keep production up.[59] And that works only if the loans keep coming, which they won't. Any fall in the production rate of shale oil that gives the appearance of a peak having been reached will only

---

59 The Red Queen is a character in Lewis Carroll's book *Through the Looking-Glass*, who famously said about a running a race: ". . . it takes all the running you can do, to keep in the same place . . ."

be a signal to the financial markets that the jig is truly up, and the collapse of the shale industry will be underway.

These conundrums of growth, debt, and energy have provoked a lot of chatter in the financial media that an epochal reset in the money system itself is in the cards. This reset will be provoked as the debt-based system implodes from either debt default or monetary inflation from massive central bank QE in response to default. One popular theory has the International Monetary Fund (IMF) stepping in as a backstop of last resort to bail out the floundering central banks with "liquidity" in the event of a breakdown in global credit obligations, or currency failures, especially in the case of a destabilized US dollar, the world's so-called reserve currency. The IMF would supposedly furnish its members with a vehicle called Special Drawing Rights (SDRs). The IMF itself can't decide exactly what this vehicle is—a kind of money, a line of credit, an international unit of account, or something else. At one point they tried to label it "an imperfect reserve asset," which doesn't tell you much. SDRs can theoretically only be used by IMF member nations to meet balance-of-payment problems and a few other intermediary functions between sovereign countries. Technically, the SDR is composed of a "basket of currencies," including the US dollar, the Euro, the British pound, the Japanese yen, and, since 2016, the Chinese renminbi.

It remains unclear what the citizens and institutions of individual nations would use for money in practical transactions in the event of a global financial crack-up—and I would define that crack-up as the loss of trust that obligations and contracts will be met with a subsequent freeze on transactions. It would certainly not be SDRs. In theory, the SDRs would supposedly support existing currencies, but that may be an idle wish since fiat currencies such as the US dollar, the euro, and the renminbi are faith-based monies backed by the presumed productive capability of each nation. But if that productive capability is impaired—say, by failing energy supplies—then no amount of SDR backup might avail, since the IMF is not a nation and has no productive capacity, and the SDR is backed by nothing but the full faith and credit of its members.

One might surmise from all this that the device of an SDR is just another manifestation of the overinvestments in complexity that threaten the excessively complex global economy. It is an abstraction made possible only by the already excessively abstract global monetary system that includes a dark netherworld trade of daisy-chained abstract derivatives with poor accountability. I suspect that when that system wobbles and fails, the IMF will fail with it, along with its abstract "imperfect reserve asset."

A seamless transition to a new global monetary regime seems unlikely. A period of anxious disorder will follow the failure of the old order, and each nation might be on its own for a period of time until it demonstrates an ability to get its own monetary house in order. A new order for settling international trade could take a long time to rebuild, and it may never be as free of impediments as it was during the current era of "globalism." Monies of different kinds and qualities would not instantly and easily cross borders, despite the wonders of computer technology—assuming, of course, that the World Wide Web endures as we know it, which may be assuming too much.

Russia and China have been busy stockpiling gold since the crash of 2008, supposedly in anticipation of debuting gold-backed currencies, or so the web chatter speculates. The US pretends to be indifferent to gold, and our own gold reserves have not been audited for decades. Economists have argued the disutility of gold-backed currencies through the ups and downs of the twenty-first century. It is certainly true that gold is mispriced in relation to the trillions of dollars in debt extant in the world and the quadrillion in unregulated derivatives. I have to conclude that any gold-backed currencies could only be introduced after a painful unwind of those obligations, which would entail the repricing of just about everything and make it obvious that official currencies have been fatally compromised by central bank interventions in the debt markets.

In the extreme scenario of a long and deep financial crash, I can imagine silver specie being reintroduced as circulating currency with gold behind it as a generally uncirculated store of value—that is, to be hoarded. In this event, there would be enormous value destruction of other "assets," including paper securities and real estate other than good farmland.

Money, after all, has three functions: a medium of exchange, a store of value, and an index of the price of things. Anything affecting to act as money must be able to do all three, and the performance must be stable over time.

If citizens lose faith in their currency, they will stop using it as a medium of exchange, and if its value is uncertain, it fails as a store of value and an index of price. Precious metals at least have one quality that other modern monies don't: no counterparty risk. If a lot of investors go broke trading in janky securities and currencies, gold and silver retain their value and utility.

We're some distance from that outcome as I write, and so another fantasy circulating in the financial press is that money, as we know it, will be replaced by national digital currencies created by computer "magic." Certainly, some countries (e.g., Sweden) are working hard to eliminate cash. People in other nations may not be so ready to submit. An unappetizing by-product of this scheme will be government's ability to track everybody's income and spending, in essence, a financial "surveillance state" with an expanded range of punishments for noncompliance. I doubt that Americans will fall for that bargain. They will no more give up cash than they will give up guns. I also doubt the utility of digital currencies—sovereign cryptos in the Bitcoin mode. They are created from nothing (besides a great deal of energy-wasting computer calculations in server farms), and they are backed by nothing. The Bitcoin experiment looked like a popped bubble in early 2019, having fallen from its 2017 high of nearly $20,000 to about $3,500. It clawed back above $12,000 in late June of 2019 and then oscillated wildly across the summer. Gross instability is the last thing you want in something vying to be a currency.

Blockchain currencies are yet another exercise in techno-narcissism. They are also another manifestation of overinvestment in complexity (and abstraction) in a system that already suffers from too much of that—to the degree that many Wall Street activities that pad the bank accounts of people already very rich are abstract rackets with no relation to the real economy. Blockchain currencies claim to be safe from hacking, but that doesn't jibe with reality. Many hacks of cryptocurrency exchanges and thefts of individual crypto "wallets" (records of ownership) have already occurred. The fact that they rely on highly synchronized computer networks ought to be of concern to people, if only for the implied fragility.

Don't forget that the internet is at the mercy of the electric grid, the world's largest machine, as it is sometimes called. The American electric grid is in notoriously shabby condition. Draw your own conclusions.

Chapter 14

POLITICS: JACOBINS AWOKENING

The 2016 election victory of Donald Trump was a tremendous shock to the political establishment. Though it is hard to separate the two, the loss of Hillary Clinton may have been a worse and more enduring shock—and an insult—to the upper orders of American society than Mr. Trump's win. It sent the Democratic Party and the entrenched managerial cohort of the federal government—sometimes called the *Deep State*—into a fugue of hysterical rage that shows little sign of letting up years later. If anything, after the midterm election of 2018, with Democrats taking a majority of seats in the House of Representatives, the rage has only ratcheted up. In late March 2019, Special Counsel Robert Mueller's final report to the attorney general ended the mystery of "Russian Collusion," finding no evidence of such crimes among the president and his associates. Mr. Mueller's testimony before two House committees in mid-summer was a tragic fiasco, showing him to be barely familiar with the particulars of his own investigation. His botched mission to rescue the reputation of the FBI and the Department of Justice only added more and greater questions about their operational legitimacy. Despite Mr. Mueller's failure, Democrats running several House committees threaten to investigate every aspect of Mr. Trump's life, career, and family, with impeachment still at the top of their agenda.

Mr. Trump's election victory was a by-the-numbers demonstration of William Strauss and Neil Howe's "Fourth Turning" thesis.[60] In the book of that title about generational cycles in history, Strauss and Howe posited that in the early twenty-first century, America would enter a dark phase of political distress as the bad choices of previous decades—pointless wars, deindustrialization, fiscal profligacy, bank bailouts, identity crusades—worked their hoodoo on the body politic. They predicted the rise of an archetypal leader they called "the Gray Champion," an elder figure from the Baby Boom generation who would "produce a sweeping political realignment" in a time of great stress and uncertainty. Few political observers would describe Mr. Trump so grandiloquently, but he certainly has been a tremendous disrupter of the established order.

History is a prankster. You order a Gray Champion, and cosmic room service sends up a casino developer and New York real estate mogul with a laughable hairdo, a big mouth, and no experience running a government. And yet, there he is, a rococo crypto-monarchist in gilded plastic trappings, living in the White House, oddly representing the most hapless segment of the electorate, the dispossessed, flyover *deplorables*, who had been cruelly ejected from a secure and comfortable middle-class existence when so much US industry was loaded on that slow boat to China. His adversaries on the Left generally, and in the Democratic Party especially, inflamed over the 2016 election result, coalesced in a hysteria-driven campaign to hound him relentlessly.

The campaign to delegitimize Mr. Trump as a candidate and then as president, which may have gotten assistance from inside Barack Obama's circle of White House advisers on behalf of Democratic Party nominee Hillary Clinton, began during the election itself. It continued at an even higher pitch when Mr. Trump won the election and humiliated Mrs. Clinton, after which it became a crusade to drive him from office. The nation's so-called "intelligence community," led by then CIA director John Brennan, geared up a scheme to spy on the Trump campaign; entrap some of its

---

60 William Strauss and Neil Howe, *The Fourth Turning: An American Prophecy—What the Cycles of History Tell Us About America's Next Rendezvous with Destiny*, New York: Broadway Books, 1997.

minor players (Carter Page, George Papadopoulos, Donald Trump, Jr.) in cloak-and-dagger operations (that backfired); spin out a Russia collusion "narrative" that painted Mr. Trump as "an agent of Putin;" and engineer the appointment of a grand inquisitor, Mr. Mueller, to launch investigations that would prepare Congress with a grand brief for the president's removal. The scheme enlisted the upper echelons of the FBI and its parent agency, the Department of Justice, the US State Department, and the British Intelligence Agency MI6, along with the major national US media, to carry out the plan. MI6 was brought on board as a work-around to the legal strictures that forbid the CIA from spying operations on US citizens.

The centerpiece was a folio of falsehoods concocted by a former MI6 operative named Christopher Steele—the "Steele Dossier"—alleging criminal connections between Mr. Trump and Russian government agents, as well as salacious sexual tales about Mr. Trump in Moscow. The Steele Dossier was financed by the Hillary Clinton campaign through the "opposition research" company Fusion GPS, with payments from the Clinton campaign laundered through the political law firm Perkins Coie, who then shepherded the dossier into the above-named federal agencies. Steele was on and off the FBI payroll during the period, but continued his activities after being discharged as a paid intel source. The dossier was distributed to a roster of select senators and congressmen (e.g., Mark Warner, D-VA, the late John McCain (R-AZ), and Adam Schiff, D-CA), as well as the *New York Times*, the *Washington Post*, the cable news networks, and other news outlets. It is unclear in late summer of 2019 how much of the material contained in the document was written and assembled by Mr. Steele or by employees of Fusion GPS (including especially Russian affairs specialist Nellie Ohr, wife of then associate deputy attorney general Bruce Ohr).

The dossier was used as the primary predicate document by the FBI in obtaining surveillance warrants against Mr. Trump's associates. It was never properly vetted under the FBI regulation known as the "Woods procedure." Emails and memoranda between the FBI and DOJ officials involved, made public since 2018, showed clearly that they knew the dossier was false but that they submitted it to the "special" FISA court judges—under the Foreign Intelligence Surveillance Act—regardless.

The machinations among the upper echelons of the DOJ, FBI, intel officers from other agencies, and a cast of political characters in and out of government carried a strong odor of criminal misconduct and led to the firings
and resignations of many, including then FBI director James Comey, deputy
director Andrew McCabe, his legal counsel Lisa Page, FBI chief of counter-
intelligence Peter Stzrok, the FBI's top lawyer James Baker, deputy attorney
general Rod Rosenstein, aforementioned Bruce Ohr, and others, with suspi-
cion cast on Mr. Brennan, former attorney general Loretta Lynch, former acting attorney general Sally Yates, and Special Counsel Robert Mueller himself.
In the deeper background of possible liability were members of Mr. Obama's
inner circle, White House National Security Advisor Susan Rice and then
UN ambassador Samantha Power, and even Mr. Obama and Mrs. Clinton.

Mr. Comey's handling of the Clinton email server case is an exceptionally tangled tale. But it is worth noting that the emails at issue go back
to Mrs. Clinton's days at the State Department and would have included
communication pertinent to the Clinton Foundation, which had been under
investigation for failing to report donations from foreign sources as far back
as the year 2001. Then deputy attorney general James Comey took over that
investigation in 2005, while Rod Rosenstein happened to have been the
head of the Department of Justice's tax division between 2001 and 2005.
Robert Mueller was director of the FBI during that period, and in 2009
was involved in the Uranium One deal that parlayed substantial amounts of
uranium from the US to Russia—and netted the Clinton Foundation more
than $150 million from Russian business interests. It was in Mr. Comey's
interest to find a way for all of that (and more) to disappear down the memory hole, and he engineered that outcome in July of 2016. His "dismissal" of a
potential case against Mrs. Clinton was a departure from regular procedure;
it is up to the attorney general to decide whether to prosecute or not, not the
FBI director. The real objective of Mr. Comey's maneuver may have been to
protect himself, Mr. Mueller, Mr. Rosenstein, and other colleagues from their
prior actions involving Mrs. Clinton and her foundation, even if it meant
damaging her presidential campaign.

The move was a fiasco because at exactly the same time other figures in
government already mentioned were working hard to advance Mrs. Clinton's

run for president, including committing actions of dubious legality to cast Mr. Trump as a Russian tool. Ironically, the scheme amounted to the Clinton campaign, the FBI, and other government actors *interfering* in the 2016 presidential election, which was exactly what the same inquisitors had blamed Mr. Trump and the Russians for.

The catch was, in spite of the exertions to nail down the election for Mrs. Clinton, she lost. And once she lost, those US government players who "meddled" in the election on her behalf stood to be exposed for the simple reason that Mr. Trump would soon be in charge of the executive branch, would appoint his own people to run the agencies, and would be in a position to discover all that misconduct. It would be reasonable to suppose that Mr. Trump would be mighty angry about what has since been termed a "soft coup" against him by his own government. It was this fear of being exposed for sedition—a serious crime—that drove the hysteria among Mr. Trump's antagonists.

And so, the second part of the crusade to delegitimize Mr. Trump spooled out after the election. The aforementioned rogue officials doubled down on the Trump-Russia collusion narrative, using their tractable friends at the *New York Times*, the *Washington Post*, and the cable TV News networks to relentlessly report new fabricated angles on the "narrative." It was basically a cover-your-ass operation, a smoke screen to divert attention from their own crimes.

Mr. Trump made a tremendous blunder in the early going of appointing Senator Jeff Sessions of Alabama as attorney general. During the transition of power after the election, Mr. Sessions had engaged in chitchat with Russian ambassador Sergey Kislyak at a buffet line on the Washington party circuit, and on that account was later importuned to recuse himself from supervising the then newly appointed special counsel, Mr. Mueller. It ought to be obvious that ambassadors from foreign lands are in Washington precisely to circulate among the political leaders of their host country and talk to them. It is normal procedure, not a crime, especially during presidential transitions. The same ploy was run on Mr. Trump's newly appointed national security advisor, General Michael Flynn, who was eventually charged with the crime of lying about a phone conversation with Russian ambassador Kislyak a few weeks before the inauguration. The idea that an incoming official of Gen. Flynn's rank—the president's national security advisor—should *not* talk to foreign ambassadors is preposterous, of course. He was nevertheless hectored into pleading guilty by Mr. Mueller, with the understanding that a courtroom defense would bankrupt the general.

Now that a new attorney general, William Barr, and his deputy, John Durham, have opened new counter-investigations into the handling of "Russiagate," there is reason to suppose that many of these government officials will find themselves in courts of law. It is already documented in testimony to House and Senate committees that the FBI and Department of Justice officials knew that the Steele Dossier was a fabrication, and that the agencies used it anyway to kick-start their campaign against Mr. Trump. Thus, the predicate for the Mueller investigation was knowingly dishonest. The primary purpose of the Mueller investigation was to distract and conceal the agencies' criminal misconduct once Hillary Clinton lost the election, and it became clear that the paper trail of emails, memoranda, and archived investigatory reports could not be dumped down the memory hole. The Mueller investigation was, in effect, a cleanup operation to protect the reputation of the agencies and of persons involved in the actions against Mr. Trump, including especially those of Mrs. Clinton and Mr. Obama.

This digest of the Russiagate scandal is a necessary prelude to a consideration of what can happen going forward from this uneasy moment in American politics. It is not an appraisal of Mr. Trump's presidency, which has been unorthodox so far in just about any way imaginable, though it has not produced the "sweeping political realignment" suggested by Strauss and Howe. However, during the Trump interregnum, the Democratic Party has shifted ever further from the center to the frontiers of the political Left. The party has been busy Jacobinizing its agenda, personnel, and mode of operation. The Jacobins were the faction in the French Revolution that became the most radicalized, seized control of the government, and conducted the persecutions known as "The Reign of Terror" (June 1793–July 1794), in which seventeen thousand of their various enemies were sent to the guillotine. The leaders of that sordid episode, Maximillien Robespierre and Louis Antoine de Saint-Just, were among the Terror's final victims when the revolution suddenly turned on them.

The French Revolution happened in the first place because of the obvious maldistribution of wealth and privilege under the *ancien régime* of monarchical France. The Seven Years' War with England (1756–1763), and then France's generous support in arms, troops, and ships for the American Revolution, functionally bankrupted the crown and thrust the nation into deep depression by the late 1780s. Crop failures in 1787 and 1788 badly aggravated the suffering of the ordinary population with bread shortages ("Let them eat cake," said the queen, Marie Antoinette). Even the silk industry collapsed. The popular revolt against the royal establishment proceeded by increments. But the attempt to create the new institutions of a republic proved extremely vexing and difficult, and the frustrations drove the squabbling factions to flights of unreason.

The Society of Jacobins was a political club that met in a former monastery of the Jacobins in the Rue Saint-Honoré in Paris.[61] Their hero was the political fantasist Jean-Jacques Rousseau (1712–1778), whose animating ideas were the natural goodness and perfectibility of man. The Jacobins were excited by the opportunity to test this philosophy in the laboratory of revolution. Their basic operating procedure—a remarkably persistent one, as shown by Lenin and many others in the modern era—was that *the ends justified the means.* They were committed to a radical restructuring of society beginning with the literal killing of the monarchy and all traditional authority with it. From there they proceeded to the destruction of traditional culture in many matters of daily life and its replacement with new institutions and manners. Despising the Catholic Church, they expropriated the Church's property, exiled thirty thousand priests, put many clerics to death, and endeavored to "de-Christianize" the nation with a new quasi-religion, the cult of the Supreme Being, that eerily resembles the New Age spiritual grab bag of today. They abolished the Sabbath, banned Christian holidays, and promoted seasonal festivals instead; got rid of the traditional calendar and replaced it with one that reset historical time to begin at the Revolution; and they changed the names of the months, and redesigned the week from a span of seven days to ten days. And much more.

---

61 The name Jacobins was given to the Dominicans in France because their first headquarters in Paris was in the Rue Saint-Jacques.

Along with Rousseau's ideas, the Jacobins were inspired and propelled by Thomas Paine's arguments in *Rights of Man*—yet they had no compunction about suspending due process, and the rule of law entirely, when it suited their purposes. In the tribunals of The Reign of Terror, the accused were disallowed the right to defend themselves or the representation of an attorney. (Robespierre himself was an attorney and should have known better.) Their bad faith in this extravaganza of bloodshed led to their swift downfall in 1794. Once Robespierre and Saint-Just were overthrown and executed, the Jacobins no longer played a role in the revolutionary government, which just drifted until Napoleon Bonaparte stepped into the picture in 1799, and then, as we know, everything changed again.

The Democratic Party post-2016 has adopted many of the Jacobins' principal beliefs and methods. When societies get into deep economic trouble, the wish arises to punish those who retain wealth and power and to confiscate their property, and to do it by force, beyond the niceties of law. We ought to worry about where this is taking the party of the Left in America today. Philosophically, they are aligned with Rousseau's idea of the perfectibility of man and a neo-Jacobin wish to engineer equality of outcome in a world of uncertain events where people are born with different abilities. That belief in perfectibility is given an additional crypto-Gnostic mystical spin with the wish to transcend nature and erase boundaries, starting with biologically determined sexual categories.

Faced with the extraordinary income inequality of the early twenty-first century—which accelerated greatly under the eight years of Mr. Obama—the Democrats are driven increasingly to redistribution schemes. Some of the new members of the US House of Representatives elected in 2018 frankly call themselves "democratic-socialists," as did presidential candidate Bernie Sanders in 2016. In the run-up to the 2020 presidential election, the candidates are already proposing higher income taxes, a guaranteed basic income, and a new confiscatory wealth tax on assets, not the *income* from assets, but a tax on what you already own.

The disgraceful racketeering in medicine and higher education that has ruined so many lives prompted the Democrats to push for nationalized health care ("Medicare for all") and free college education, despite the fact that the USA is functionally bankrupt as much as Louis the XVI's government was in 1789. But the time for grandiose, top-down bureaucratic reform is over in a new era where the collateral of *future industrial economic growth* is absent and therefore further debt accumulation is foreclosed without destroying the value of money. Even theoretically, the money to pay for those programs doesn't exist.

Those systems as currently evolved are probably unreformable. They need to collapse and emergently remake themselves at a much leaner and smaller scale. And they will, whether we go with the flow of that process or swim as hard as we can against the current.

In the meantime, as financial markets wobble and the global economic system groans, Mr. Trump will be left holding the bag. He and the Republican Party, which he took hostage in 2016, will be blamed for the economic distress that follows. They will not be able to wriggle out of it. As presidents go, Mr. Trump is pretty old. A financial train wreck could take a toll on him, physically and psychologically. Even if he remains healthy, the Republicans might look elsewhere for a nominee in the 2020 election. But the party will have been successfully "Hooverized," handing the Democrats a victory, despite the unreality of their policy positions—that is, if the nation manages to avoid a deeper institutional breakdown that would disable the election process itself.[62]

In 2021, a new unified Democratic majority regime could take the reins of government in an economic crisis worse than the Great Depression of the 1930s. Their *ends justify the means* modus operandi—in which systematic dishonesty is the means and wealth confiscation the end—will not avail against the tides of widespread bankruptcy and socio-economic dysfunction. With the Republican Party mortally wounded and *hors de combat*, it will be the Democrats' turn to disgrace themselves by making the economic hardship even worse, spending the country into deeper bankruptcy, and destroying the value of the US dollar in the process. By the mid-2020s, both parties may be history, and conditions in the nation could bring shocking changes in the way we are governed.

---

62 Hooverized: viz, Herbert Hoover, thirty-first president, 1929–1933, blamed by many for the Great Depression.

One probability is a loss of faith in the federal government per se, and the rising perception of its impotence among a broad range of the public, both Left and Right. Polls show politicians are the *least* trusted among all professional occupations. The paralysis of Congress is obvious to everyone, but the other functions of government are equally immobilized. Extant antitrust law and securities regulation go unenforced. Immigration policy is incoherent. States and big cities defy federal law with their own "sanctuary" policies, and the feds don't challenge them. The federal government can't even figure out how to coordinate old US laws on marijuana with legalization in the states where it has become a big business—with hefty contributions to state *and federal* taxes. One thing the federal government seems to carry out sedulously is using the National Security Agency (NSA) to spy on its own citizens by collecting every email, tweet, and retail purchase record sent over the internet.

If trust breaks down further, there could be a years-long struggle to run the government by other means. By these I'm referring to military administration, a shifting leadership from new coalescing factions fighting the equivalent of a low-grade civil war, the emergence of new parties and movements that galvanize the public's desperation, and perhaps the rise of a would-be American Bonaparte (or a cornpone Hitler). Even a determined autocrat may not be able to rescue the government's credibility, given the problem of national insolvency, failing energy supplies, and declining industry. The upshot will be a federal government that cannot discharge its duties and obligations. The states may not fare much better. In such a scenario, the public will be forced to reorganize their affairs at the local level. It will be obvious by then that in a crisis of permanent economic contraction, all the other colossally overscaled systems in American life—retail commerce, banking, agriculture, medicine, education, etc.—have lost the ability to function, just as the government has, and that harsh passage will be the opportunity for Americans to rebuild their local economies and communities at a scale consistent with reality, especially the new reality of resource and capital scarcities.

In that future, we'll discover that the end of techno-industrial economic growth does not mean the end of economic activity. "Growth" as represented in statistical GDP reports is not a productive activity in and of itself. A post-collapse disposition of things will mean a society of people working diligently at activities aimed at carrying on civilized life on a smaller scale, by other means than fossil-fuel-powered, technological hyper-complexity. In fact, so much has to change and be reorganized on the ground that there will be no end of the work to be done by the hardy individuals who can persevere through tough times.

It's too early to know how they will get paid, or what they will use for money, or what the new political boundaries will look like, but it is not necessarily the end of the world, just the end of a phase of history. Considering the immense entropic blowback we live in now, and the cognitive dissonance it throws off, releasing the throttle on modernity's awful entropy engine will mean a more purposeful and meaningful life for those who get through the upheavals to come.

Chapter 15

CULTURAL NOTES: FUMBLING TOWARD KAFKA'S CASTLE

olitics, the saying goes, *is downstream from culture*. Whatever you find distasteful, offensive, or insane in American politics can be traced back to our culture: the age-old improvisational business of manners, protocols, rules, rewards, and punishments that society constructs to enable its own functioning within the context of time and place. A casual survey around America these days reveals shocking degrees of neuroticism, delusion, dishonesty, and functional failure in culture. The result is a political dwelling place that looks more and more like Kafka's Castle, a techno-bureaucratic update of a particular kind of solipsistic Hell.

The economically stranded former working class has devolved into a tribe of tattooed savages sunk in anomie. Those a rung up in the *middle* middle class are not far behind them, as vocations and incomes disappear, debts mount, and desperation creeps over the scene. Obesity and its by-product, diabetes, run at record levels thanks to "innovations" in the food-processing industry (another racket), and the absence of other traditional social satisfactions that have been destroyed by television and smartphones. Opiate addiction and suicide are the new normal in the flyover states. The triumphant completion of suburbia has produced yawning ugliness on the landscape, an epidemic of loneliness, family dysfunction, and a dismal cavalcade of mass shootings in public schools. Meanwhile, the upper rungs of society are enfogged in a contrived obsessive moral panic over race and sexual relations.

Can a people recover from such an excursion into unreality? The USA's sojourn into an alternative universe of the mind accelerated sharply after Wall Street nearly detonated the global financial system in 2008. That debacle was only one manifestation of an array of accumulating threats to the postmodern order, including the burdens of empire, onerous global debt, population overshoot, fracturing globalism, worries about energy, disruptive technologies, ecological havoc, and the specter of climate change.

The sense of gathering crisis persists. It is systemic and existential. It calls into question our ability to carry on "normal" life much further into this century, and all the anxiety that attends it is hard for the public to process. Disinformation rules. There is no coherent consensus about what is happening and no coherent proposals to do anything about it. Bad ideas flourish in this nutrient medium of unresolved crisis. Lately, they dominate the scene on every side. A species of wishful thinking that resembles a primitive cargo cult grips the technocratic class, awaiting magical rescue remedies to extend the regime of Happy Motoring, consumerism, and suburbia that make up the armature of "normal" life in the USA. The political Right seeks to *Make America Great Again*, as though we might return to the booming postwar economy of 1955 by wishing hard enough.

The thinking class, meanwhile, squanders its waking hours on a quixotic campaign to destroy every remnant of an American common culture and, by extension, a reviled Western civilization they blame for the failure to establish a heaven on earth of rainbows and unicorns. By the logic of the day, "inclusion" and "diversity" are achieved by forbidding the transmission of ideas, shutting down debate, and creating new racially segregated college dorms. The universities beget a class of what Nassim Taleb prankishly called "intellectuals yet idiots,"[63] hierophants trafficking in fads and falsehoods, conveyed in esoteric jargon larded with psychobabble in support of a therapeutic crypto-Gnostic crusade bent on transforming human nature to fit the wished-for template of a world where anything goes. In fact, they have only

---

63   Nassim Nicholas Taleb, "The Intellectual Yet Idiot," *Incerto*, September 16, 2016, https://medium.com/incerto/the-intellectual-yet-idiot-13211e2d0577#.pbipdn1dg.

produced a new intellectual despotism worthy of Stalin, Mao Zedong, and Pol Pot.

In case you haven't been paying attention to the hijinks on campus—the attacks on reason, fairness, and common decency, the kangaroo courts, diversity tribunals, assaults on public speech and speakers themselves, the denunciation of science—here is the key takeaway: It's not about ideas or ideologies anymore. Instead, it's purely about the pleasures of coercion, of pushing other people around, of telling them what to think and how to act. Coercion is fun and exciting! In fact, it's intoxicating. It is endorsed by the authorities (the faculty chairs and college administrators), echoed in top organs of the media like the *New York Times* and NBC News, and in the products of Hollywood—and rewarded with brownie points and career advancement.

The new and false idea that something labeled "hate speech" is equivalent to violence floated out of the graduate schools on a toxic cloud of intellectual hysteria concocted in the laboratory of so-called "post-structuralist" philosophy, where sundry body parts of Karl Marx, Herbert Marcuse, Michel Foucault, Jacques Derrida, Judith Butler, and Gilles Deleuze were sewn onto a brain comprised of one-third each Thomas Hobbes, Saul Alinsky, and Tupac Shakur to create a perfect Frankenstein monster of thought. It all boiled down to the proposition that the will to power negated all and any other human drives and values, in particular the search for truth. In this scheme, all human relations were reduced to a *dramatis personae* of the oppressed and their oppressors, the former generally "people of color," women, and the "sexually nonconforming," all subjugated by whites, especially white heterosexual males. Tactical moves among these self-described "oppressed" and "marginalized" are based on the familiar cynical credo *the ends justify the means* favored by political utopians.

This is the recipe for what we call *identity politics*, the main thrust of which these days, the quest for "social justice," is to present a suit against male white privilege and, shall we say, the horse it rode in on (Western civ). A peculiar feature of the social justice agenda is the wish to erect strict boundaries around racial identities while erasing behavioral, sexual, and ethical boundaries. Since so much of this thought monster is actually promulgated by white college professors and administrators and white political activists, against people like themselves, the motives in this concerted campaign might appear puzzling to the casual observer.

I would account for it as the psychological displacement of shame, disappointment, and despair over the outcome of the civil rights campaign that started in the 1960s and formed the core of present-day progressive ideology. The civil rights campaign was an imperative consequence of America's victory in the Second World War over manifest evil. After proclaiming ourselves *the beacon of hope for the Free World*, the USA was compelled to dismantle the armature of racial segregation—or lose its moral standing. It was also simply the right thing to do. It changed a great deal about everyday life in the nation, but, alas, it did not bring about the hoped-for golden age of racial comity.

It should be obvious that so much of the legislated efforts to assist black America didn't work out as expected and instead produced tragic unintended consequences, starting with the income distribution policies that promoted single-mother parenting and the absence of fathers in the lives of their children. The racial divide in America is starker now than ever, even after a two-term black president. Today, there is more grievance and resentment, and less hope for a better future, than when Martin Luther King Jr. made the case for brotherhood on the steps of the Lincoln Memorial in 1963. More recent flashpoints of racial conflict—Ferguson, the Dallas police ambush, the Charleston church massacre, Charlottesville, etc.—don't have to be rehashed in detail here to make the point that there is a great deal of ill feeling throughout the land, and quite a bit of acting out on both sides.

The black underclass is larger, more dysfunctional, and more alienated than it was in the 1960s. My theory again, for what it's worth, is that the civil rights legislation of the early 1960s, which removed legal barriers to full participation in national life, induced considerable anxiety among black citizens, who did not feel altogether comfortable with the new arrangements, for one reason or another. And that is exactly why a black separatism movement arose as an alternative at that particular time, led initially by such charismatic figures as Malcolm X and Stokely Carmichael. Some of the acting out around all that was arguably a product of the same youthful energy that drove the rest of the sixties counterculture: adolescent rebellion. But the "Black Power" movement was essentially a call to remain segregated from, and frankly hostile to, the majority white culture that had just ended the Jim Crow era. The residue of the movement persists today in the widespread ambivalence about making covenant with a common culture for all Americans, and it has only been exacerbated by the now long-running "multiculture and diversity" crusade that effectively nullifies the older concept of a national common culture altogether.

What follows from these dynamics is the deflection of all ideas that don't feed a narrative of power relations between oppressors and victims, with the self-identified victims ever more eager to exercise power to coerce, punish, and humiliate their oppressors, the "privileged," who condescend to be abused to a shockingly masochistic degree. Nobody stands up to this organized ceremonial nonsense for fear of inviting opprobrium and "cancellation." The punishments are too severe, including the loss of livelihood, status, reputation, and advancement, especially in the university. Once branded a "racist," you're done. And venturing to join the disingenuous and oft-called-for "honest conversation about race" is certain to invite that fate.

Globalization has acted, meanwhile, as a great leveler. It destroyed what was left of the working class, a.k.a. the lower middle class, which included a great many white Americans who used to be able to support families with their labor. Hung out to dry economically, this class of whites fell into many of the same behaviors as the poor blacks before them: out-of-wedlock births, absent fathers, drug abuse, crime. Then the Great Financial Crisis of 2008 wiped up the floor with the *middle* middle class above them, foreclosing on their homes, livelihoods, and futures. In their desperation, many of these people became Trump voters—he saw that the white middle class had come to identify themselves as yet another victim group, allowing him to pose as their champion.

Identity politics is a racket no less than the scams in Wall Street, so-called health care, the college loan industry, and all the other dishonest games running in this troubled moment of our national life. Ultimately, it defeats the idea that we are individuals, with the awesome responsibility to take care of our own lives within the framework of agreed-upon standards of decent behavior. Adopting a reality-optional ideology is not a way to thrive in this challenging world undergoing deep change.

The evolving matrix of rackets that prompted the 2008 economic debacle has only grown more elaborate and craven as the old *economy of stuff* died and was replaced by the *financialized economy of swindles and frauds.* Almost nothing in America's financial life is on the level anymore. Life in this milieu of immersive dishonesty drives citizens beyond cynicism to an even more desperate state of mind. The suffering public ends up having no idea what is really going on, what is actually happening. The toolkit of the Enlightenment—reason, empiricism—doesn't work very well in this socioeconomic hall of mirrors, so all that baggage is discarded for the idea that reality is just a social construct—just whatever story you feel like telling about it—and what you report your feelings to be. On the Right, Karl Rove expressed this point of view some years ago when he bragged (of the George W. Bush White House) that "we make our own reality," and the Left says nearly the same thing in the post-structuralist, narcissist malarkey of academia: *Your "lived experience" is your truth.* In the end, both sides are left with a lot of bad feelings and the belief that only raw power has meaning—which beats a path to unconstrained wickedness.

Erasing psychological boundaries is a dangerous thing. When the nation's economic rackets finally come to grief—as they must because their operations don't add up—the American people will find themselves in even more distress than they've endured so far. Erasing behavioral boundaries—a culture of *anything goes and nothing matters*—ends up nullifying the social contract as a general proposition. The bankruptcy of the nation will be complete, and nothing will work anymore, including getting enough to eat.

Historically, public hysterias eventually run their course. As author and blogger Jasun Horsley puts it, we are in a "liminal" situation, "a time in history when ambiguity and disorientation have assumed epic proportions."[64] The end of an empire is exactly that liminal moment, made worse by the end of the economy that enabled it. The people struggle to find who to blame

---

64 Jasun Horsley, "What is Liminality?" http://kunstler.com/what-is-liminality-guest-essay
-by-jasun-horsley/; and also https://auticulture.com/.

and what to do. One thing after another obstructs their ability to carry on, to make a living, to find satisfactory relationships, to protect their offspring, to find purpose and meaning in daily existence—until merely being in this world seems like the worst sort of swindle. "Now what . . . ?" becomes the universal cry of anguish. What will happen next? What will you do? Better figure something out, because no one else will figure it out for you.

PERSONAL CODA

I've been following the story of the imploding American scene for years in a series of nonfiction books and the four-novel series I wrote under the *World Made by Hand* banner about a post-collapse American future. In the midst of all that, seven years ago, I moved from the upstate tourist town of Saratoga Springs, New York, to a much smaller old factory village fifteen miles east on the other side of the Hudson River. Saratoga was a good place to live, one of the few still-functioning Main Street towns in the upper Hudson Valley. I'd been happy there for over thirty years. After getting divorced in 2004, I rented a series of places and sat out the worst of the housing bubble, watching and waiting. But property prices in Saratoga had gone way up after 9/11, as a lot of Wall Street hedge-funder types vamoosed from New York City looking for a safe place to raise their kids. I couldn't afford anything there by then. So, I had to make other arrangements.

In 2011, I wanted to set up a homesteading operation for myself. I searched for a place to buy out here in the village and found one literally at the edge of town, on a little dead-end lane eleven feet outside the village tax district. You could walk to Main Street in a few minutes. That was important to me. I didn't want to be marooned out in the boonies, having lived through oil crises in 1973 and 1979 when folks were having fistfights on the gasoline lines and you never knew whether you'd be able to get from point A to point B. I was also not looking to become a farmer on a large acreage. That's not something you start in your sixties. I wanted to have a big garden, plant fruit trees and berry bushes, raise chickens, and have a small woodlot to heat the house in the event that fossil fuel was not available for the furnace.

The property I found on three acres had a house that was only a few years old. The price was good compared to the Saratoga scene. It had been built by a couple who both had jobs in the commercial construction industry, lost them after the housing bubble popped, and had to sell their dream house. I felt sad about the deal for their sake, but they were going to sell it one way or another, so they might as well sell it to me. The husband was very generous as the sale went forward. He walked me through all the systems and the appliances and explained how everything worked: the heating setup, the septic, the well-water purification apparatus, and the bathroom ventilators. I'd looked at a lot of other houses in this town, most of them much older, with layer upon layer of old incomprehensible plumbing and electric runs and basements that looked like crypts out of a Vincent Price movie. The utility room in this place was so clean you could have a picnic in there, and all the pipes were clearly labeled. I also liked the way the kitchen and the living room flowed into each other, with the fireplace at one end—because you can never get people out of the kitchen at a party—and it's been a long-held principle of mine that the best way to keep a social network alive is to throw parties at frequent intervals.

The real estate closing in the bank boardroom was a sad episode, with a lot of tears shed, and I learned that the couple split up a few weeks later. It happened that I was getting very sick just then from a janky hip replacement back in 2003 that had given me cobalt and chromium poisoning, and over the next couple of years I had to make a lot of visits to the blood lab at the local health center, where the wife of that couple worked as the phlebotomist. She was very nice to me, considering.

I'd moved in just before Christmas and was very busy that first year despite feeling physically ill. The property was rectangular, with a cleared acre between two one-acre wooded lots. The house was in the middle of the clearing. One of the first things I did in the spring was to hire a guy to put up a deer fence around the cleared acre. It was imperative because the deer population here in Washington County, New York, is completely out of hand—young people don't go hunting much anymore, they play video games instead—and that first winter of 2012 I had a herd of seven deer practically living on the front porch. It was like a petting zoo out there some mornings. I never could have grown any food with them hanging around. The deer fence was *Jurassic Park* grade: steel with twelve-foot posts pounded four feet into the ground. The guy who put it up was a state trooper with a side business who styled himself as "The Post Pounder."

While that was going on, I laid out a large, formal kitchen garden with a picket fence of black locust posts (extremely rot resistant) and cedar pickets around it to keep out the rabbits. I had to reinforce it with steel chicken wire around the bottom. I put the garden gate on an axis to the front door of the house so I could just march out and grab stuff for dinner. In April, I began planting the first twelve of about twenty-five fruit trees in the back of the house. They were a mix of apples, cherries, and pears. I would add plum trees in the years to come and more Asian pears. Then I got the berry bushes in: raspberries, gooseberries, black and red currants, and hardy Siberian kiwis. I planted grapes along the south-facing run of deer fence, with steel wire strung between the posts to train them on. I added a row of hazelnuts between the berries and the fruit trees. After that, I built a structure called a hoop house, a plastic covered half-pipe, ten by eighteen feet, on the north side of the garden so I could extend the growing season for table greens. I added a three-bay compost bin, also made of long-lasting black locust wood. Finally, I built a carpenter Gothic chicken house and painted it green with purple trim.

What I learned is that it is not so easy to grow food. The first couple of years I had little trouble from pests. Then the cabbage moths and other insects discovered what I was up to here, and it's been a pitched battle ever since, though it turns out there are plenty of ways to keep them down without spraying on chemical poisons. I had to reinforce the garden fence some more to keep the woodchucks out. (I also got a Havahart live trap and am now running a lively woodchuck/opossum deportation program.) Still, I get all the lettuces and kale I need (and then some), enough potatoes to get through the whole winter, plenty of peppers, tomatoes, eggplants, cucumbers, squashes, leeks, and onions. I have a big herb bed at the center of the garden, and a separate mint bed a hundred feet away, because mint spreads like crazy on underground runners and you don't want it anywhere near the garden. It takes about ten hours a week to attend to things around here. My original scheme included grassy paths between the planting beds, but I eventually got rid of the grass and put down wood chips. The planting-bed scheme is highly geometrical. Order with nature induces serenity. The soil up here is "boney" (as they say), full of rocks, but years of composting has helped the planting beds a lot, and the waste from the chicken house goes into the compost assembly line to enrich it.

The fruit trees are just coming into production now at seven years—though the Asian pears bore prodigiously from year two. I might have gotten more apples last year, but we had a frost in early May during the few days that the trees bloomed, which kept the bees down, and very little fruit got *set*. The people who built the house planted one peach tree on the property. Over the years, I've had two bountiful crops of peaches and several years of no peaches at all. The second big peach crop (2017) was so huge that several big limbs split off under the weight of the peaches. The lesson there is to pinch off some of the fruit as soon as it starts developing (duh . . .).

The weather has been erratic over the past seven years. We had two winters with practically no snow, one with consistently warm temperatures, and several with blizzard after blizzard and deep freezes. The second year after I planted fruit trees, we had no rain in August and September, and I had to run soaker lines into the orchard. Last August and September, it seemed to rain every day. The gooseberries and currants produce well and reliably, and I've been making a lot of jam over the years. My grapes are a bust, though I was careful to acquire hardy varieties suited to upstate New York. Some of them just up and died, and the rest seem to struggle year after year. I don't know why. But wild grapes have started climbing up the deer fence from the outside, and they're doing very nicely. I discovered a wild mulberry tree growing just outside the fence, too. The Siberian kiwis grow on vines (male and female), and they are twining on the fence very nicely, but I haven't seen a single kiwifruit yet. The five hazelnut shrubs are growing well, too, but so far I have gotten a crop of exactly one nut in all the seasons I've been here. That nut is on my desk as I write. My nut crop.

Raising chickens is pretty easy. I started with a flock of five Ameraucanas—Gladys, Mabel, Pearl, Doris, and Solange—and I stick with that breed, with occasional off-breed adoptions from friends forced to "re-home" their birds. They run from rich dark brown to silvery gray, and they lay blue eggs. On summer days it gives me great pleasure to see them ranging busily around the property, enjoying dust baths and chasing cabbage moths. But over the years I have lost many to predators, especially in springtime when the beasts of prey have new litters to feed. The first flock is long gone now, and I keep rotating new hens in. I've actually interrupted a couple of attacks. I watched a red-tailed hawk swoop down and snatch one of the girls, and when I shrieked at it, *mirabile ictum*, the hawk dropped the hen. She recovered, only to be killed in a raccoon attack later that year. Last September I heard some panicky screeching while I was on the phone in the kitchen and rushed outside to see a red fox about to chomp down on an orphan red hen I had acquired from a friend who gave up on chickens. The fox was just lying in the grass, holding Little Red between his paws. When I rushed onto the scene, the fox dashed off, but Little Red was never the same and died a few weeks later. It's been a hard lesson to learn that that's just how nature rolls. The fox was amazingly beautiful and was just following its genetic program. I wouldn't try to kill it, but I might set the Havahart with some hamburger this year and take the fox on a long ride across a couple of rivers. One afternoon, a dog from down in the village got in the homestead when I carelessly left the front gate of the deer fence open. He slaughtered two of my birds. I located his owner a few hours later. The dog was a big hound named Leroy. His owner said he was exceptionally stupid. For all that heartache, I get a regular supply of the lovely blue eggs (inside, they look just like any white or brown egg).

The village at the bottom of my hill was really beat when I got here in late 2011, and things actually got worse since then. There were four or five factories here in the nineteenth and twentieth centuries and they're all gone now. Steel plows were forged here for the settlers busting sod in the Midwest in the 1840s and '50s. Other factories started out making linen textiles from locally grown flax but shifted to paper products later. The Battenkill River, a tributary of the Hudson, runs through the village in a big loop. It's a renowned trout stream, but it also provided water power for nascent industry, and later on, hydroelectric power. Two of the former factory sites still have small hydroelectric outfits running on them, but the power is just sold into the grid and the factories are gone. I can imagine a future when the village might have to rely on local hydroelectric power.

A number of ruins from the industrial age still stand around town. There's a set of dressed stone bridge piers from the early railroad (1869) standing mutely in the river on the south side of town. They are partially collapsed now, with trees sprouting out of them. The trestle they supported is long gone. Last summer I went swimming in the Gardon River in France under a Roman aqueduct that still spans the stream after nearly two thousand years. It amazes me how quickly the landscape has swallowed up the residue of industrial America, and how brief that heyday was. The ruins of the Dunbarton Mill (1879–1952) stand a quarter mile downstream from the old railroad trestle piers. The final product of the Dunbarton Mill was toilet paper, a sad irony for a town that has kind of gone to shit. Only a shell of the redbrick main works remains, along with the cut stone millraces that directed the river flow to the factory's power train. The ingenuity embedded in all that astounds me. The steam shovel had barely come into use back then, and most of the work was accomplished by gangs of men with hand tools. Now the site is considered a toxic brownfield. The owners of the derelict property are nowhere to be found (considering the outstanding tax and cleanup liabilities). The village has no idea what to do with it.

A short walk up the riverbank from there stand the ruins of the town's railroad infrastructure. There's a six-track freight train marshaling yard, several abandoned warehouses, a forsaken coal elevator, and a station house. The decrepitating ensemble speaks eloquently about what the nation so recklessly and tragically threw away. The railroad connected with the industrial city of Troy, New York, thirty miles south, and from there, you could connect with passenger trains to Boston and New York City. The journey from the village down to Manhattan took five hours altogether in the 1920s, including the change of trains. The drive today takes at least four hours—of intense concentration behind the wheel—sometimes longer in terrifying weather, with tractor-trailers bum-rushing you in the passing lane. I'm not persuaded that it's an improvement in the overall quality of life. By the early 1900s, an interurban electric trolley system also connected the village to Glens Falls, Saratoga, Lake George, and as far north as Warrensburg in the Adirondacks. It's been gone since before World War II. Now it's hard to imagine a country so well arranged. (You can still come across the running mound of overgrown right-of-way out in the woods.)

The population of the village runs about 2,500. As I write, two businesses are closing, accounting for about one hundred jobs. One is in a storefront on Main Street that repackaged pharmaceuticals in daily dose plastic bags for old folks. The New York State regulators shut it down. It was rumored that some of the employees were drug addicts. The other business was the Kmart, whose parent company, Sears, has been whirling around the drain for years. The loss of the Kmart is perhaps unfortunate because it was the only "variety" store within twenty miles, and it employed a lot of people in town. But I consider it a harbinger of better things to come. If we're ever going to reactivate our Main Streets, it will probably take the death of these national chain stores—which is very much underway, as their business model is failing. Most people assume that the chain stores' business will be replaced by internet shopping, but I believe that notion is in error. Amazon and other online retailers have a problem: every single item they sell has to make a truck journey to the customer. That model assumes, among other things, that the trucking industry will endure. I don't believe that for a moment, knowing what I do about the oil predicament and the poor chance that we will electrify trucking. And we have missed the window of opportunity for rehabilitating the US railroad system. In a coming era of capital scarcity, the money just won't be there. We blew it. Boo hoo.

I think a lot these days about getting a couple of mules and training them to the cart. Perhaps you think that sounds crazy. I'll just leave it at that.

Strange to relate, I'm optimistic about the town's future. It's surrounded by good agricultural land that is completing its journey out of agri-biz-style dairy farming, which was ruinous to both the land and the farmers who practiced it. New ventures are well underway here for growing food and raising livestock at a different scale. The Battenkill River remains a reliable source of water power, and the Hudson River is just five miles away, with its connection to New York City and the Atlantic Ocean in one direction and to the Saint Lawrence River via the Champlain Canal going north. The landscape of tender hills and hollows is ceaselessly lovely.

The good news (for me) is that my health returned after about ten years of travail. The cobalt poisoning induced a supernatural kind of fatigue in me, as though somebody had turned up Earth's gravity two clicks on the dial. My legs felt like I was lugging around a couple of Yule logs. I couldn't work physically more than an hour without needing a rest. My hands and feet were numb. I had weird cramps and other bizarre sensations. I used to lie in bed at night with that old Doors song "The End" playing over and over in my head, a real *dark night of the soul* situation. Still, I managed to get a lot done. Strange to relate, I wrote more books in that ten-year period than in any previous decade—at the same time that my book advances were going down, down, down, like so many other people whose vocation is writing books. I even managed to remain mostly cheerful, despite what was happening to my body, not to mention my professional focus on writing about economic collapse. It just didn't drag me down. It strangely energized me.

Meanwhile, I had a bunch of surgeries. And then, something happened. I began to realize that I was feeling a lot better physically. I returned to a normal relationship with gravity. The strange neurological sensations faded away. I could work outside longer, shoveling and lifting. I was able to sleep through the night after years of soul-crushing insomnia. I had recovered mysteriously from what I thought was the last roundup. I couldn't feel more fortunate. I couldn't feel more grateful for being here, in this astounding world—even if it's in trouble and we humans are in trouble with it. Of course, something else could happen to me and take me out of here, and surely will at some point, as it will to you, too.

That's the real brotherhood and sisterhood of this journey we're on. We all have the same destination. You are not alone! What a spectacle it is! History is rolling out. Roll with it! Be kind to each other. Be true to yourselves. Put your shoulders to the wheel! Carry on, bravely!

INDEX

West Texas Intermediate (WTI), 207, 226

Wheat Belly (Davis), 155

When Trucks Stop Running (Friedemann), 211

Whidbey Island, Washington, 95–96

white nationalism
  as facet of identity politics, 113–114
  Freeman on, 115, 123–130

white supremacy, as racism, 85

Wickett, Josh (pseudonym), 73–94
  background of, 74–82, 88–91
  on City of Baltimore, 92–93
  code of conduct of, 84–88
  and Freddie Gray incident, 91–92
  on the future, 93–94
  importance of music to, 79–80, 84, 89, 93
  on military life, 76–81
  on need for "American" culture, 73–74
  on nonprofits, 90–91
  and Open Works community center, 82–83

Wilcove, David, 216

Windham County, Connecticut (the Quiet Corner), 113, 116

Windows on the World, 81

wind power, 23, 29–35, 102

wind turbins, 30–31

The Wire, 75

wishful thinking, 3–4, 244

working class, 243, 247

World Made By Hand series (Kunstler), 83, 251

WTI (West Texas Intermediate), 207, 226

X

Xi Jinping, 198

Y

Yanukovych, Viktor, 37

Yates, Sally, 234

Z

Zahm, Luke, 54

Zubkova, Anna, 122

Zuckerberg, Mark, 42